Number Three: Tarleton State University Southwestern Studies in the Humanities

WILLIAM T. PILKINGTON, *Series Editor*

Cannibals and Condos

CANNIBALS
AND CONDOS
Texans and Texas along
the Gulf Coast

by Robert Lee Maril

 Texas A&M University Press
College Station

Library of Congress Cataloging-in-Publication Data

Maril, Robert Lee.
  Cannibals and condos.

  (Tarleton State University southwestern studies
in the humanities; no. 3)
    1. Gulf Region (Tex.)—Description and travel.
  2. Man—Influence on nature—Texas—Gulf Region.
  3. Maril, Robert Lee—Journeys—Texas—Gulf Region.
  4. Gulf Region (Tex.)—Social life and customs.
  I. Title. II. Series.
  F392.G9M37 1986   976.4′1   86-5759
  ISBN 0-89096-276-6

Manufactured in the United States of America
FIRST EDITION

For Travis and Lauren

Contents

Preface

Usually it takes my son and me ten minutes to get from my parked car, behind the dunes, to the beach, fifty yards beyond. He pushes his two-year-old body through sand and brush in determined strides, muttering to himself, "Beach, beach." When we come to places in the dunes where he might get stuck, he raises his hands up to me to be carried. I tell him, "Beach, Travis, beach," drawing out the words and making them last. Under the hot dune air his eyes get bigger, a smile appears. His face glows rosy in the midday sun.

We cross the dune line, and I put him, shoes first, in the ankle-deep sand. He sees the gulls and starts for them in a toddler trot, arms thrust forward and swinging against his naked sides. He gets another few steps before the gulls see him, take off, circle, and drop down a little beyond their first spot. Travis stops, looks back at me for bearings, then is off again in hot pursuit.

When he has "disappeared" all the gulls, then only does he head for the surf. Fearlessly the young Texan marches into the waves, oblivious to the water temperature, the ocean bottom, the purple jellyfish strewn about like used balloons. He sees nothing, I suspect, but a vast set of waves and bubbles and friendly noises waiting to play with him.

Raised in the suburbs of Oklahoma City, the closest I got to a beach for the first decade of my life was a small fresh-water lake with a mud bottom and trucked-in sand. The sand stretched a few feet to the clubhouse, where we could buy hamburgers from Labor Day to Memorial Day. The place was called the Branding Iron. I thought all water was chlorinated.

The summer I turned thirteen we flew to San Francisco. My father spent three days at his medical convention while the rest of us—my mother, older brother, younger sister, and I—did Chinatown, Fisherman's Wharf, and the rest. All I remember from that first visit to San Francisco is that it was cold and rainy in July.

When the convention was over my father rented a car and we headed for Los Angeles. The plan, concocted by the travel agent, was to follow the thick yellow line on the Triple A map straight south, spend a few days at Disneyland, then fly back to the hot summer waiting in Oklahoma. We got as far as Carmel. My parents walked us through the downtown, my mother trying to get at least one of us interested in the galleries, then over to the public beach.

We spent the rest of the afternoon there, more or less entranced. By some unspoken agreement my father left, only to return with a key to a cottage not two blocks away. For the next six days my brother and I perfected our sand castles. Our goal was to construct one so tough it would last through the night. We had mixed success. My mother made picnic lunches of peanut butter and jelly, grapes, apples, and cokes in glass bottles. The next to the last night we were there, my little sister got sick and vomited on me. It was the last time we slept in the same bed. After nearly a week we drove on to Disneyland, spent a day bored at the rides, then flew back home. It was 1961. My brother still has some slides, mostly of San Francisco.

My idea of a beach for the next eight years was a Carmel beach. Yellow moist sand and weathered cypress, misty surf slamming against cliffs with houses clinging tenaciously to them. When I saw my first Texas beach one college spring break, I was disappointed. Later, when my wife and I moved to South Texas, I kept comparing South Padre Island to Carmel or Bar Harbor or even Coney Island.

The question here, usually raised only in whispers by Texans fearing to seem disloyal, is whether Texas beach, in fact the entire Texas Coast, stretching from Beaumont–Port Arthur to Brownsville, is not simply ugly, a wasteland, best ignored by most Texans and certainly scorned by those outside the state. I think a fair answer is to ignore comparisons to the East and West coasts. Obviously the lands bordering the bays and the barrier islands facing the Gulf of Mexico are quite different from Carmel or Martha's Vineyard. If we carry around in our heads an image of what coastal areas *should* look and be like, then Rockport and Galveston will be poor cousins, indeed, to many spots on the Atlantic and Pacific coasts.

To be truthful, the most outstanding characteristic of our coastline is its flatness. Flat as a pancake. For miles and miles. Anyone who approaches, say, South Padre Island by car, expecting, at the last minute, to see a Texas Big Sur, is in for a letdown. No doubt about it: the beach is flat. Even the oldest and biggest dunes, some rising fifty feet high, are not really that big a deal, except for the fact that they are surrounded by land that resembles a tabletop.

The immensity of the Texas Coast also astounds the eye. The coast goes on and on, in most places unbroken by man-made structures. This immensity, I think, intimidates many people. As a species we like to see examples of our handiwork even when we want to get away from our large cities. Along the coast it is possible to go mile after mile and see no signs of any human presence.

The Texas Coast impresses the visitor with much the same feeling of hugeness as does the Big Bend. The comparison may seem odd, but the point is that both geographical areas can overwhelm the senses by size alone. Big Bend, of course, overwhelms us with its mountains, deserts, and weird rock formations. The coast overwhelms the eye with what is not there. The coast is grasses, salt flats, irregular and unplanned bays, a surf line that extends north and south as far as the eye can see, and too much sky.

Much of the coast is physically isolated as well. With the exception of the Houston metroplex, Beaumont–Port Arthur, Corpus Christi, and a few other areas, all that dot the shoreline are small fishing villages and, more recently, resort developments. A Texan has got to want to go to Sea Drift or Bay City to get there. He cannot pass through on his way somewhere else, because the communities are dead ends. The Gulf, in its way, physically isolates much of the underpopulated coast as much as the surrounding hundreds of miles of ranchland and desert isolate Big Bend.

What we have, looked at one way, is an immense, flat, isolated strip of land unlike any stereotype of what a beach should be. To boot, it borders an even more immense, more unknown liquid mass we know as the Gulf of Mexico. There are many who have an instinctive fear of the sea. I'm one of them. We can usually take it in small doses, especially if it is controlled. We like quiet seascapes; a cold fear comes over us when we hear that yet another suburb in California has slipped into the Pacific or that sixty-foot racing yachts are gobbled up off England in seventy-foot seas. Not to mention what goes on under the water, where our imagination can run rampant.

Seen in its own right, the Texas Coast possesses a unique beauty. It is not the beauty of the Swiss Alps, or of Cape Cod. It requires patience and the development of an appreciation for what is there, not a pining for what is not. The coast requires time, and in our society time is in short supply.

Time is one thing I've been lucky enough to have. Over the last decade I have walked the shores, fished the bays, and lingered in coastal fishing towns. I have begun to penetrate this region's façade. The Texas Coast is much more than it would at first appear to be. That's not to say that it appeals to everyone, only that the person who accepts superficial first impressions can miss much that is worthwhile.

What follows in these pages is an exploration of Texans and the Texas Coast, an exploration informed in equal parts by my experiences as professional researcher, traveler, and father. This modest search is an attempt to cast off cultural myths and misconceptions that may blind us to what is important about these lands. There is something to be learned from all the peoples who have inhabited this area, including those indigenous Indians often referred to as cannibals. Instructive, too, are the ways in which some coastal communities have been developed within the last ten years. Condos now predominate in towns where the houses of commercial fishermen once stood.

I focus on these two symbols of the Texas Coast, cannibals and condos, to emphasize the considerable historical changes that have taken place in a very short time. Yet man is not the only player along the Texas Coast; nature has always held the trump card. All peoples of this region, be they clad in loincloths or bikinis, have adapted to the physical constraints of a land that borders a sea. They have depended on the natural resources that sustain life, maintain their culture, and change at the whim of season. Everyone, whether alleged cannibal or oyster dilettante, adapts to the hurricanes and storms that periodically ravage the area, randomly clearing it of human edifices and rearranging geography.

Within this thematic framework I examine, albeit briefly, a variety of issues and policies that affect the Texas Coast. I raise questions unasked or long since forgotten: how we live along the ocean, how we both appreciate and neglect it, how we protect ourselves from it, and how our various levels of government manage it. Rather than limit this exploration of a land and its people to antiseptic statistics or impersonal debates on specific issues, I include my own family's concessions and adaptations to some of these same basic concerns.

Travis does not leave the beach voluntarily. Even though he has run down his share of sea gulls, jumped his quota of waves, covered his body with sand and shell fragments, and seen "fishes" in the warm tidal pools, he is not—is never—ready to go home. I have worked out a method to withdraw him more-or-less peacefully from the beach. I tell him that it is time to go, that the beach is tired. We say, in unison, my son and I, "Goodbye, beach. We'll be back another day."

He believes, with the heart of a young child, our mutual pronouncement. I am not at all sure. The Texas Coast and its peoples have gone through more change in the last ten years than in the previous one thousand. While the region remains in large part a frontier, parts of the Texas marshes, bays, and sands are truly tired. We have pushed and are pushing these lands and waters beyond our own abilities to use them, and beyond their capacity to recover. Too, we are bringing rapid changes to the way of life of many coastal inhabitants. Change is not always welcome.

My hope is that we can limit and control how we develop these lands, so that they are, in the process, not totally destroyed. There is no inevitability in what we choose to do, only the knowledge that certain choices lead to certain consequences. From the beginning it should be understood that I am not against development, against business, against condos—I am simply against destruction of the environment in the name of development. Neither would I return to a time before the Spanish explorers first appeared on Texas shores.

My hope, too, is that new uses of these lands will not totally displace those who depend on them for their livelihood. We need not render jobless all commercial fishermen and others who depend on the coast. Nor should we judge their ways of life to be outmoded—legislate them away—and, through "management," manage their undoing.

Fortunately, there are a number of reasonable positions between these two extremes, between cannibals and condos. This book attempts to define some of these alternatives by focusing on concerns grounded in my own growing understanding of these Texans who are born, are raised, work, begin their own families, and eventually die on lands that border a sea.

Acknowledgments

This book would not have been possible without the help of many people. I would like to thank Michael V. Miller, Lorry King, and Dennis Berthold, each of whom commented on earlier drafts of this project. Davor Jedlicka was particularly helpful in providing data on the Texas Coast. Conversations with Bill Kulevsky originally helped stimulate my interest in particular coastal topics. Discussions with Valerie Gunter were also productive. At Texas Southmost College, Denise Joseph and Tony Zavaleta supplied historical information on indigenous coastal Indians. Juliet Garcia provided the administrative support to complete this effort.

I would like to give special thanks to those at the Texas A&M University Sea Grant Program, in particular Laura Colunga, Allen Martin, Bill Clark, Lorry King, and Feenan Jennings, all of whom have provided a wealth of information to me over the years about the Texas Coast. A grant from that program facilitated this work, but all facts and opinions expressed herein are mine and no one else's.

I would also like to thank all those who have taken the time to share their experiences and knowledge of the Texas Coast with me. In particular, many residents of Texas coastal communities have been especially generous.

Last, this book would have not been possible without the editorial skills, moral support, and direction of Andrea Fisher Maril.

Cannibals and Condos

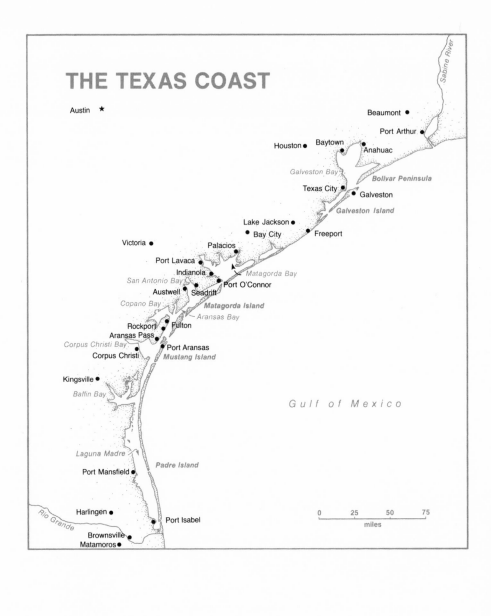

THE TEXAS COAST

Austin ★

Beaumont ●

Port Arthur ●

Houston ● Baytown ●

Anahuac

Galveston Bay

Bolivar Peninsula

Texas City ●

Galveston ●

Galveston Island

Lake Jackson ●

Bay City ● Freeport ●

Victoria ●

Palacios ●

Port Lavaca ●

Indianola ● *Matagorda Bay*

San Antonio Bay Port O'Connor ●

Austwell ● Seadrift ●

Copano Bay *Matagorda Island*

Aransas Bay

Rockport ● Fulton ●

Aransas Pass ●

Corpus Christi Bay Port Aransas ●

Corpus Christi ● *Mustang Island*

Kingsville ●

Baffin Bay

Gulf of Mexico

Laguna Madre

Padre Island

Port Mansfield ●

Rio Grande

Harlingen ●

Port Isabel ●

Brownsville ●
Matamoros ●

0   25   50   75

miles

Sabine River

Mucking It, Then Afterwards the Enchiladas

I crouched in muck. It oozed against my rubber boots and sucked at my poncho. I could see absolutely nothing in the predawn dark. With one arm I cradled the 12-gauge; with the other I protected my face from the rain. I had long since thrown my soggy hat out into the shadows. It was raining, not heavily, but enough to get the point across. It was also cold.

To ward off the coastal chills, I had on a T-shirt, ski sweater, and jacket, all in theory protected by the borrowed poncho. In practice, of course, none of the above did much more than serve as a sponge for the rain that whipped against me, soaking into folds of the poncho, dousing my jacket and sweater. The T-shirt, my favorite, initialed with a national brand of running shoes, was the last line of defense. I swore that when the T-shirt fell to the assault I would stand up and walk back along the levee the mile and a half in muck and knee-deep water. There I would abide in exile in the truck until my friends transported the remains of my body to a dry and warm place.

I was here by choice, so I could leave anytime. I should have thought the whole thing out more clearly, of course, before I came along. I should have spent all of the night before in preparation. The selection of gear should have been made as if my life depended on it. I would now give, for instance, two hundred dollars in cash for some hip waders, the kind that have reinforced toes and adjustable suspenders. With that kind of gear I could walk through an arctic swamp and never feel a thing.

As it was, water drooled down my forehead, aiming for my eyes, and slime inched its way up my gun barrel. It could not be too much longer until dawn. I wondered if any of my friends had frozen to death. I pictured the sun rising majestically behind the rain clouds to reveal long lines of low-flying Canadian geese headed directly toward me. As taught, I would leap to my feet, blasting them out of the sky. Four or five goose bodies would drop around me. Turning, I would glimpse in the frozen ooze the remains of my ponchoed friends, their bodies bound to the earth by ice crystals. They would all have smiles on their frozen countenances, having glimpsed the goose parade at rosy-fingered dawn just before the pleasant side effects of freezing to death ended their hunting careers.

The geese did, in fact, appear. We heard them long before sighting them. Their noise roused the blood in my limbs. Honking to each other in the low-slung rain clouds, they invaded the silent dawn. As they grew nearer, the beat of their wings drove me crazy with the excitement of anticipation. I could easily visualize the leader breaking a path through the clouds, calling to his friends, looking for a good place to feed before continuing on the long journey south.

A voice to my left said, "Not yet. Wait." We waited. The geese were now directly overhead. The same voice yelled "Now!" I jumped up, slipped on my poncho, and fell to my knees, jammed the butt of my Sears special into the ooze to extricate myself, and was deafened in both ears by shotgun blasts on both sides of me. I regained my feet, got some wind and rain directly in my face and down one ear (the sensation of which compares to hearing fingernails drawn across a blackboard), sighted at a shadow in the sky, and blasted away. Thankfully my gun still was operational.

I blasted through the legal number of shells, then popped in more from the bandolier around my waist. Now I shot at shadow tails and distant honking, hoping for a lucky shot.

The line of geese never broke stride. We stood, the five of us, looking like mud wrestlers in camouflage, not a goose corpse within ten miles of us.

One friend and I left the others for new fields a hundred yards to the southeast. We disguised ourselves as mud and waited. Twenty minutes later a new flock of unsuspecting gooseflesh headed our way. Again, we could hear them long before establishing visual contact. My T-shirt caved in to the elements. Only my thumping heart, excited by the coming kill, kept me warm. At the signal I jumped to my feet and pulled the trigger.

The sound of pellets thunking against flying goosebreast is tantalizing. I know for certain I hit at least one of the birds. I jammed in a few more rounds and shot at empty air. Again, not a bird in sight for all our efforts. Did they wear armored vests, like presidents?

Once more I went through the by-now-familiar routine. Again the geese flew on toward Belize, undaunted. My heart cooled down; I noticed my toes were numb and that the rain, for no apparent reason, had finally stopped.

We slogged back to the truck. The long, low, wide-open fields of sorghum and cotton stubble lay saturated in the fall rains. Five miles to the east lay the Gulf, turning frigid under autumn skies. The one-lane roads that bordered the fields had become narrow streams with occasional holes made for stumbling. At times the water was too deep for my high rubber boots, and I was forced to stop, retrace my steps, and find a shallower route.

Toward the southeast the gunfire continued. The only other sounds were the sloshing of boots against water, then the popping of boots removed from muck. Occasionally we stopped to rest. It was a tiring walk and was made more so by the 12-gauge, which became increasingly heavy in my hands.

Eventually we joined up with the others, deciding to give up on the geese and hunt any stray ducks that, in the early morning hours, could be caught unawares. We drove around and around, eventually sighting some birds landing on a small pond a half-mile from the paved road. Two of us were dropped off, and the others motored to a spot a few hundred yards away; from there they would take a ten-minute hike to the pond and flush the birds from cover.

According to this plan, my friend and I would get a clear shot at the ducks as they passed overhead. Again we trudged through the muck, in a hurry to be in hiding before the first shots would set the ducks to flight. Halfway to cover we heard the blasts. Three mallards zoomed by, out of range by a good hundred yards.

So we trudged back to the truck. The father was still yelling at his son for firing too soon, well before they themselves were in range. The twelve-year-old looked glum. The father turned away in anger. "Why did you start shooting?" I asked the boy. "I couldn't help it," he said. Indeed he could not. Twice more that morning the kid tried shooting ducks from two hundred yards away. He lacked the patience and maturity to wait. So he fired from the hip, gunslinger style, at anything that moved. For my own safety, I stayed well behind him the rest of the day. After each episode his father yelled at him. I made a mental note never to yell at my own son when we both stood with loaded shotguns and not five feet between us.

Eventually we ran into a Parks and Wildlife ranger. He was immaculately dressed, despite the weather. Only a speck of muck had established a beachhead on his shiny black boots. I was impressed with his friendliness, his efficiency—and the huge cannon he wore on his hip. He checked our stamps and our shotguns and sent us on our way.

By then it was past ten in the morning. Minus the father and son, we drove in to Rio Hondo, parked, and spent time cleaning ourselves as best we could. The restaurant was almost empty when we entered. I slumped into a booth, ordered coffee and enchiladas, and hoped the cook would hurry. She did. The enchiladas were hot and juicy, with thick slabs of melted cheese on top. I gulped them down, adding large amounts of hot sauce, then mopped up the plate with some pieces of flour tortilla. I then chased the food down with two cups of steaming coffee. We sat around the table telling stories about the hunt and laughing about the father and son.

Two hours later I was at home soaking in a hot bath, drinking more coffee and swearing out loud that I would never hunt geese or ducks again. Of course I would, probably the next week. But for the moment I enjoyed the swearing and the feel of the hot water against aching joints. I stayed in there long past the time when the water remained tepid despite my attempts to reheat it.

I hosed down my borrowed poncho, pants, boots, and other paraphernalia. The muck, diluted, covered my driveway and flower bed with a rich black silt. I took a wet rag to my shotgun, removing the worst of the muck, then cleaned out the bore with patches and remover from my kit. I oiled the 12-gauge, inside and out, and slid it into its case in my closet. I stored away the bandolier after removing and cleaning each shell. Only then did I feel that the hunt was really over.

That season, my first, I killed three ducks. Not a lot, but a start. In spite of several sustained attempts on my part, the Canadian goose population never considered me a serious threat. By the season's end I had begun to master the art of staying warm in the muck and ooze, but the geese continued to elude me. Lukewarm in body, I would jump to my well-practiced feet, shooting in vain at the high-flying honker silhouettes.

In fact I began to enjoy the pre-ambush more than the shooting. Lying in the slush among the trampled grasses, trudging through the dark along the soggy roads—all this was so totally unnecessary that there was a wisdom to it. I found that I liked to be up and around when everyone else was asleep in bed. The talk of the ducks before dawn, the hum of geese coming down from the clouds for a look-see, the bull-frog throbs in what the biologists have labeled the "wetlands," these are sounds that many never experience.

I have, as yet, not tasted wild goose from my own table, but I hear that it is very good. In the meantime, perhaps forever, spicy enchiladas with hot, hot coffee will suffice after a frigid morning in the coastal muck.

Yogi the Rigger

Yogi the rigger worked harder than any other man I have ever seen. He was all over the rear deck of the shrimp boat, working the nets, shinning up the outriggers, changing the oil filters, taking his turn at the wheel, all the while talking a blue streak and smoking cartons of Marlboroughs.

Every afternoon about 4:30 he would get up from his nap and cook the crew dinner. His food tasted of his travels. He lived in Brownsville, but he was from Florida and Louisiana by way of Detroit. His meals were covered in strong, thick, southern gravies with just a hint of Cajun spice in them. He could bake hamburger and make it taste special, and when the fresh food in the freezer was gone he would cook up fish, squid, and crabs he caught from the nets.

And of course he cooked shrimp. He picked them carefully from the plastic laundry baskets where they were tossed after being deheaded and before being iced in the hold. He would pick out a handful, talk about why one was too large, another was colored wrong, and a third just did not feel right to him. Then he would take his bouquet to the kitchen refrigerator, proclaiming that he would personally dehead anyone he caught messing with his shrimp.

Later, he would simmer the shrimp in boiling water, throw in salt, mumble his theories on exactly how long fresh shrimp should be boiled, and tell a fish-eating story for good measure. He had five basic fish-eating stories, each a variation on how he single-handedly barbecued enough fish for his whole neighborhood. Then he would take a store-bought cocktail sauce, add huge amounts of horseradish tempered with lime and salt, and serve it all up to the crew. Or he would go the fried route but skip the gravy.

I sopped gravy for three weeks and learned to hunger for it. The craving was especially strong when the first smells hit me as I lay dozing in my bunk, fighting the summer heat with Yogi's little electric fan that never worked for longer than half an hour.

The only food I ever turned down from Yogi was the squid. He made his usual incantations over it, but I just could not stomach the taste, more or less like fried cardboard. Yogi said he was disappointed in me.

Yogi worked with an enthusiasm that was infectious. He was a perfectionist at the nets. He had his method and never varied from it, regardless of weather conditions or time of night. Occasionally he would tell me he felt lazy that day but somebody had to do the job, and do it right, so off he would trudge. He set the nets, cleaned them, sewed them shut, untangled them, whatever it took. He would follow behind Larry, the header, cleaning up what Larry had chosen to miss. He hated a job done poorly.

He would not comment on the captain's decision to fish one place, skip another. We did not get many boxes of shrimp on that trip, or at least not as many as the other trawlers were bringing in. Yogi would look over at me with hangdog eyes, mouth pouting under a long droopy moustache, and give a big wink. He just would not believe the captain could be missing all those shrimp that lay there on the bottom just waiting for a Texas shrimper to come by. Later, when Yogi took the wheel at three in the morning, he would mutter to me about the injustice of it all, how the captain had the boat but Yogi had the brains and the experience. He just could not stand a job done poorly.

The happiest I ever saw Yogi in those three weeks some fifty miles off Houston Bay was when we ran into the whites. They were large, going maybe seven or eight a pound in the stores, and would go for five dollars and up at the dock. We hit them one morning about ten o'clock just as we were about to clean up the rear work deck. Yogi went crazy. He kept saying over and over to no one in particular, "Goddamn, goddamn, goddamn." He headed like a madman, throwing the huge shrimp heads one way, their tails another. When he had finished filling one basket full of tails, he ran his hands through the silvery crustaceans as if sifting through gold dust.

We kept catching whites for another six hours, and Yogi kept heading as if possessed. I do not know how he kept at it so long. I took breaks every thirty minutes, and still my lower back throbbed from the stoop labor. Larry had two years of experience at heading but needed a cigarette break at least once an hour. The captain took frequent coffee breaks. Yogi never stopped. When the whites finally played out, he stood straight up from his stool, threw off his gloves in disgust, and headed for his bunk. He told me in passing that if he were captain, we would be in whites for the next three days.

One night Yogi and I swapped life histories. The moon was full, and there was a fifteen- to twenty-knot breeze coming out of the south-east. It was another balmy night in the middle of the Gulf, and the only life in sight was a trawler working two miles upwind. We had one long talking spree, and after that I considered him a friend.

He claimed he had been born and raised in Detroit and had led a normal life until his wife was killed in a car wreck. Said he was halfway through college then. He quit, joined the army, and went to Nam for two years, where he managed to keep from getting killed. He was fiercely patriotic. He could not believe the job they were doing over there. He got discharged and somehow ended up in Florida, learned how to shrimp there on a friend's boat, worked five years until the shrimp began to play out, then eventually found himself in Morgan City, Louisiana. From there he could tell the real money in shrimping was being made in Texas, so he tried Port Isabel and Port Lavaca and finally decided Aransas Pass was more to his liking. He told me four or five times that if his wife had not been killed he would be doing something else now. "No future in shrimping," he said.

There were too many gaps in Yogi's story, but I did not want to pry. A major inconsistency was how he could work as hard as he did, get paid for his labor, and still have almost nothing to show for it. He rented a trailer to live in, borrowed a car or took taxis when he was in town, and could stuff all his possessions in an old seaman's bag.

I gave him a phone number when we docked. Three weeks later I got a call from him. Said he had been back from another trip two days, and did I want to meet him for a drink? I did.

The bar was a converted gas station near the harbor. It had a pool table where the hydraulic lift had been and six tables with checkered plastic tablecloths. He must have been in the john when I came in. I sat down, looked around as inconspicuously as possible, and ordered a beer. Across the room were a few other shrimpers talking and a bar-maid watching the black and white over the bar. The floor was the original concrete. The compressor on the air conditioner thumped in every few minutes.

Yogi appeared. He had traded his standard work shirt and cutoffs for a brand-new green Hawaiian shirt and black cotton pants. The pants were too tight on him. The floral pattern in the shirt cast a pallid reflection on his neck and chin. He had a woman on either arm, but from a distance I could not tell if he was escorting them to my table or they were carrying him in my direction.

He sat down, said hello in a very low voice, and introduced the women. They sat on either side of him, occasionally getting up to get beers or talk to the barmaid.

Yogi announced to the room that the guy he was talking with had been an average header and a good friend. I asked him how he was doing. He said he had been back a few days and still had some money to get rid of. He told me the women were average, but what could you expect these days? He offered me one of them. I declined. I tried a few more bits of conversation then gave up and concentrated on my beer.

It was the same Yogi. The same carefully greased and combed hair, lying flat against his ears. Ruddy skin, varnished neck, biceps bulging, paunch squeezing out between the leaves on his shirt. But the long, droopy mustache looked even droopier. The eyes were bloodshot, looking out through slits. He was working hard on being sloppy drunk.

He took out a big wad of bills with his left hand and peeled off two on top. He handed them both to Kathy, the woman on his left, telling her he wanted another beer and some cigarettes. She disappeared. Then he rose and stumbled off to the bathroom again. That left me with Julie, who sat, legs carefully crossed, one hand on her cigarette, the other nursing her beer. Her eyes looked through me. Social custom demanded some conversation on my part.

I thought of a few things, discarded them, came out with "How long you known Yogi?" No response. "He's a helluva shrimper." No response.

Yogi returned. He now looked totally passive, physically drained, but still able to walk. He had three more days in port before the next trip. He showed me his wad of money. I told him I had to go. He nodded. His women hung close by his side.

I got out of there and drove back to my house. Yogi was really two men, but I did not see the second until we had landed. He could work or he could play, nothing in between. When he worked, he was a likeable demon. When he played, he was a drunk.

I still remember how he looked at the bar: part pirate, part stand-up comic. Yogi just did not know what to do with himself when he was not working on the back of some shrimp trawler, smack in the middle of the black Gulf, pulling, pushing, cleaning, lifting, sweating, waiting for those shrimp.

By Air

I almost missed the flight. I got fouled up at the gift shop in Love Field trying to buy Dallas Cowboys T-shirts for my kids. The woman behind the counter, who was extremely apologetic, could not get an okay on my Visa card. I was stuck, the urge to grab my card and run for the gate pulling me one way, the faces of my four-year-old, Travis, and baby girl, Lauren, pulling the other. I should have planned it better, bought them presents the day before. But I had not, and as a result I was going to miss my plane. The woman behind the counter, still apologetic, dialed the toll number one more time.

This time she got through to wherever Visa operators sit and count the credit we all accumulate. She gave them my number but skipped a digit, and they asked her to repeat it. She repeated it. The sweat began popping out on my forehead. I hate being late for anything; in fact, I would rather just not show up than be late. The numbers of the card once miscommunicated, her confusion in repeating them escalated. Finally she hung up, apologetic, finished the paperwork on the credit card, and wished me a good flight. I grabbed the T-shirts, stuffed them in my flight bag, and ran.

The problem with Southwest Airlines (more so in the early days) is that invariably the flights leave on time. I clomped through the airport in my cowboy boots, around passengers, up the escalators, and through the security check. Half a city block later I found the gate, checked in, then sat in the waiting area watching the football game on television. Strange that the flight should be so empty, only a few other people on the other side of the room, talking, smoking, commenting on the game. Another late passenger ran up to the counter, breathless, had her forms identified, then rushed past me through the door. I watched her disappear down the corridor. They must already be on the plane.

I grabbed my bags and gifts, staggered after her, and finally climbed on board. I had forgotten that Southwest Airlines usually manages to leave on time by herding its pasengers on board at least ten minutes in advance. It used to be a free-for-all in the waiting rooms when a flight was called. Old people and small children got crushed or shoved aside in the fight to see who could board the plane first and win the choicest seats. I have played that game before, and although I do not enjoy it, I do enjoy the sense of relief that comes when I am belted in, not in a middle-of-the-row seat and not at the tail end of the plane, directly across from the restrooms. When the plane crashes, the tail with four rows of seats always shears away from the main fuselage and falls into a swamp. So I have played the waiting game, positioned myself at the opportune moment next to the door that is unlocked by the ticket agent, and rushed past those who have stood in place for half an hour. I hate to fly. Fear minimizes shame.

Because the flight to Houston was only half full, I once again avoided the tail section. I dozed, ignoring the offered drinks and compulsory bag of peanuts, and we landed at Houston Intercontinental without a hitch. Almost everyone got off the plane. I moved my bags to the very front of the plane, starboard side, where the seats are two to a row and the leg room ample. I pulled out a map of Texas, threw my briefcase down on my seat to establish territoriality, and did what they imply only the insane do: I left the plane between flights.

I had not got twenty feet down the portable hallway when the rush of new passengers, those allowed to go on first because of small children or a physical disability, turned a corner and headed my way. I retreated, confused, torn between finding the captain and retaining my prime piece of real estate at the front of the plane.

As I crossed back from the corridor to the threshold of the plane, the copilot appeared. I approached him, map in hand. I have never talked to the men who fly the plane, although like everyone else I have listened for years to their stream-of-consciousness ramblings over the tiny intercom. I wondered, briefly, if he would think me a potential terrorist or some other kind of crazy. For all I know it might be against the law to talk to the crew before a flight. In any case, I asked him to show me the plan for the flight to Harlingen. He was very friendly; I do not think he had talked directly to many passengers before. He took my map, pointed out the coastal landmarks, explained the one possible alternate route, then handed it back.

I thanked him and edged past the stewardess, explaining that I was an "old" passenger and therefore had no boarding pass. Then I confronted the woman who was on my prime real estate. She seemed a little confused and preoccupied, but at my suggestion she moved quickly across the aisle. Then she rearranged some of her belongings in the seat next to me. That seemed peculiar. Suddenly two ticket agents appeared, huffing and puffing, carrying between them a man who was at least partially paralyzed. As they placed him carefully in the aisle seat next to me, the woman belted him in.

I'd forced the companion of a paraplegic to move so that I could view the Texas coast line from thirty-two thousand feet. I felt like a Jerry Lewis jerk. I told the man now seated next to me that I was sorry for the inconvenience. He could not talk. He stared straight ahead at the partition the entire flight. The woman talked to him in soothing tones from across the aisle. I looked out the window, damning planes and Southwest Airlines to eternal hell.

The plane taxied out to the runways and, after a brief wait, took off. Climbing out of the Houston smog, the plane banked and headed almost due south. Off to one side I could see, however briefly, the Houston Ship Channel surrounded by row after row of white oil tanks. From the angle and height of the plane I could not actually see the water in the channel, just the mile after mile of tanks leading off to the west. The oil tanks looked like dirty egg shells.

According to the copilot, the plane was headed for Palacios, some eighty miles to the south, on the northern end of Matagorda Bay. The land between is lush and green, rivered, with only scattered houses and farms. The plane continued to climb. Country roads and an occasional four-lane made artificial trapezoids out of the geography.

I was not sure where we were until we traversed the coastline and the broad expanse of Gulf blue water shone under the plane. With one eye on the map, the other on the land and sea below, I finally located us slightly south of Palacios. I found myself, with some surprise, looking down at one and the same time on Palacios, Port O'Connor, and Port Lavaca. The sight stunned me. I had driven through the area before, always impressed by the physical isolation of the small communities. The smaller bays and inlets that wedge into the coastal mainland off Matagorda force the highways to take a circuitous route. From the

air Palacios, Collegeport, Port Alto, Olivia, Point Comfort, Port Lavaca, Magnolia Beach, Indianola, Port O'Connor, and even Seadrift and
Austwell on San Antonio Bay all seem right on top of each other, tiny
communities huddled near or on the beach, overwhelmed by the size
of the Gulf of Mexico.

We passed quickly over the old air strip on the northern tip of
Matagorda Island. Amid the grasses and sand the geometric design of
the runway was out of place. Recently the site has gained some notori-
ety: space-age entrepreneurs have used the old air strip as a launching
pad, as yet without any major success, for the first privately owned
rocket and satellite.

From the air Matagorda Island and San Jose Island, which fronts
Fulton and Rockport, appeared as wafer-thin strips of sand dividing
the deep blue from the shallow blue and greens of the bays and inlets.
Beyond the island surf lines, the sea waters gradually deepened into a
blue that stretched to the horizon. Following it the other way, from
horizon to islands, was a jolting experience. With the exception of a
few oil rigs and shrimp boats closer in to shore, nothing broke the mag-
nitude of the Gulf, which dwarfed the island sands and the tiny bays.
The islands protect the mainlands from winds, water, and erosion, but
from the air they appear remarkably frail when matched against the
size and power of the Gulf.

We continued along the coastline almost due south, past Aransas
Pass and Port Aransas, and before I recognized it, we had almost
cleared Corpus Christi. From a great height Corpus is a town scantily
flung around a shallow bay; it appeared more water than substance,
stretched to its limit along the south and west sides of Corpus Christi
Bay. I located the bridge that links Corpus to Portland and traced the
highway back to what had to be North Shore Drive. But again, it was
the bay that dominated the scene, the western finger of the bay appearing to stretch almost to Odem.

South of Corpus the fingers and inlets of Bafin Bay squirmed like
some giant amoeba, unfettered by the restraints of civilization. There
were no roads, no buildings, nothing to be seen on the mainland
except the broad backdrop of brown brushlands. Padre Island was
peopled at its northern tip, but condos quickly gave way to nothing
more than sand and three lines of surf.

I concentrated on the patterns etched by the trawlers in the Gulf waters. At this height the white wakes of the shrimp boats were more visible than the vessels themselves. The boats appeared in groups of three, four, and five. Occasionally I spotted a lone trawler, away from the small packs.

Suddenly clouds drifted in below us, patches of white gas that blocked the Texas trawlers from sight. Then the boats were visible again, giving a depth of perception to the scene that made me aware, for the first time this flight, that I was riding in a bullet-shaped piece of metal five miles up and fifty miles out over the Gulf of Mexico. My stomach did a flip-flop off the high dive. I refocused on the inside of the plane and the color coordinated partition in front of me. Sneaking a look out of my left eye I noticed that the paraplegic had fallen asleep; petty guilt seeped out of my bones onto the cabin floor.

Inside the plane, life hummed and droned by. The captain gave his unintelligible monologue, the stews pushed their beverages, and most of the passengers slept, read, ate, or talked. I had made this trip many times before but always had ignored what went on miles below me. I usually slept the brief hour away or concentrated on some unimportant fact of my daily existence. When I had looked below, it was with initial interest but almost instant boredom. A map made all the difference. The spits of sand and water, overlaid by bits of Texas civilization, put humanity in its physical context in a way that can never be gained by seeing this same area from a car, on foot, or under sail.

Barrier islands, bays, wetlands, rivers, surfs—all are masses in transition, fixed between solids and liquids, between the solid brushland ten miles inland and the tremendous weight of Gulf water bashing against the barrier islands during the winter months or gingerly clawing at the beaches during the summer. Always, by definition, the coast is in transition, in a struggle between land and sea. The media for this transition are sand, mud, and grass. They are our fragile bulwarks, our fortifications against inundation. From the air it looks as though the Gulf must prevail.

I was the last off the plane. At the door a blast of hot, humid Harlingen air greeted me. Andrea, my wife, Travis, and Lauren were late meeting me. When I saw them drive up to the small airport (its new addition brings its gates to a total of four), I rushed across the street to them with bags and T-shirt gifts. I hugged Andrea close, but my children were asleep in their car seats. I threw my possessions into the trunk and slipped into the driver's seat. Headed south to Brownsville, I found the loop that leads to Highway 77 and became, instantly, a small dot on a geometric design.

Waiting for a Hurricane

The waiting began in earnest that night. We waited, then we waited some more. We were ready for the storm, Andrea (pregnant with Travis), our foster-daughter Belinda, our old dog, and I. But no storm arrived. We waited. Hurricane Allen, billed for two weeks in the national media as the largest Gulf storm of the century, decided not to visit Brownsville that particular Friday night in early August. Instead, he stood one hundred fifty miles off the coast, indecisive, fierce, already famous.

I had started boarding up the house around 10:30 that morning. Gathering up all the scraps I could find in the garage, I had grabbed my hammer and a two-pound bag of nails. First the windows in front, six of them. I figured the front part of the house was the most vulnerable. After I had boarded up the six windows, I crawled over the air-conditioning ducts in the attic to reach the two small dormer windows that face the street. The frame around the tiny panes in the east window was soft and fragile. I tried reducing the impact of the hammer against the nails. It did not work. One of the panes fell outward, dropping twelve feet to the flower bed below.

That is when I first felt the cool tickle of panic reach my stomach and limbs, a gradual numbing of sensation. Crouched between rafters, I felt totally helpless. If the attic windows could not stand the soft blows of nails into the plywood meant to reinforce the window frame, how could they withstand the assault of the weather report's 150-mile-per-hour winds? If the windows blew in, the rain would saturate the ceiling of our living room, all plaster and puffy insulation. And if the ceiling caved in?

The only thing to do was to hammer. I hammered for the next eight hours, nailing up a crazy-quilt defense against the storm. When I ran out of plywood I used quarter-inch imitation wood paneling. It was cheap, ineffective, but better than nothing.

My greatest achievement, almost approaching the level of art, was the bulwark I created across the back of the house. French doors cover the width of the house; across them I constructed a maze of chinks of plywood, scrap from a demolished staircase, and the remainder of the paneling. With breaks for water and weather reports, that project alone took me three hours. Every couple of feet I constructed a contingency defense, so that if one piece of wood were blown away, the wooden concoction could still protect the french doors and their many delicate panes.

As the boards went up, the house got increasingly dark. When I had finished and inspected my work I realized that all natural light had been shut out. I could no longer watch the hurricane from inside my house; I would be forced to listen to it.

I was all motion and few thoughts or words. More wood here, nails there, marshaling supplies, alternative plans if the backyard should flood, a brief discussion in monosyllables with Andrea about what to do with the animals. It was hard to stand in one place for longer than five seconds. Fear of the storm weighted my body down when I was not in motion. I felt as if the rational part of my brain had been shut down and the nonrational part, the one responsible for species survival, demanded all the oxygen and other nutrients available.

I took a final survey of the outside of the house. The trailer and boat were tucked away in the garage along with one of the cars. In haste to force the trailer into too small a space I had run it over four nails sticking out of a stray board. Now the boat and trailer leaned heavily to the left, immovable. The second car sat in a high place in the backyard, away from trees. I covered it as best I could with a tarp.

I picked up all the paraphernalia that had been in the backyard and would not fit in the boat-filled garage. The back room of our house now resembled a rummage sale: rake, shovel, hoses, spare bricks for the flower beds, sprinklers, a hoe I had never seen before, two saw horses left by the carpenters, two ladders, both of which were broken, sickly household plants exiled to the backyard to recover before being allowed back in the house, and so on. With the flotsam of a household now rearranged, I became aware of how many different objects we owned and had some responsibility for. Not to mention the house, our first one. Maybe if it had been our second or third I would have felt differently.

Inside, extra batteries were lined up for flashlights and radio.

They stood next to the Coleman fuel canisters and the stove itself, all on the bar we never used as a bar except during an occasional party. We had matches, candles, five five-gallon bottles of drinking water, plus the potential to fill our bathtubs if it really got bad. Of course I kept telling myself it wasn't going to get really bad. Maybe a little tough, like a few trees falling on the house, but not really bad bad.

All this action and that of my family was played against the background of a neighborhood thudding away at plywood and beams, where pairs of tender hands were sprouting blisters and smashing thumbs, where bodies were motivated by adrenaline flowing from the reptile side of the brain, the side that tells us, in nonwords, "When the big storm comes, all that really counts is who survives."

I kept a close eye on the reports that kept coming across the tube. Every hour or so I walked out in the front yard to look at the sky. No rain, high and obscure clouds, a very light wind. The neighborhood was still digging in; the hammer thuds continued into the early morning hours.

Twice during the night I awoke, a voice in my head yelling at me to rise, check on the storm, make sure everyone was safe and sound. Again the same reports on the television. Allen was stalled, landfall was predicted for sometime Saturday afternoon, but hurricanes, they kept saying, were fickle. There was conflicting information on each of the two local channels. I went back to bed and a light sleep.

The rains began just before sunrise. At first they were just another late summer shower. But no sun appeared, the clouds kept rolling in, these thicker and much lower, and the wind picked up. By noon the winds were gusting up to forty miles per hour and the rains came in sheets at a sixty-degree angle.

Our dog seemed oblivious to the impending storm, though I noticed he stayed out of the way, in the background, as if he knew we were too busy to give him much attention. Finally he barked to be let out. I opened the side door, let him hang his red shaggy head out the door to get an idea of what the rain was like, then pulled him back inside. Nonplussed, he demanded to go outside again. I reopened the door and backed him outside, holding onto his chain collar. I had visions of his being swept up in a gust and blown away forever. He looked at me as if I were crazy, but he lifted his leg, completed his routine, and let me pull him back in again.

The winds continued to increase. The rains came and went, each time returning stronger than the time before. By late Saturday afternoon they were blowing almost horizontally down the street. We wedged towels around the front door to soak up the water being forced through the cracks. We waited.

Periodically I shook on my white shrimper boots and, with shovel and hammer in hand, took a tour outside. I dug little trenches in the ground to ease the flooding, especially in our backyard and driveway, and hammered an occasional board. I ducked around the corners of the house to avoid the full force of wind and rain. I thought I was being clever. My fear had subsided.

Drying off inside, I then surveyed the attic and dormer windows. They were leaking, but not badly. I pushed one to test its strength and the whole window fell out and landed on the poinsettias below. The wind slapped into my face, and instantly I was drenched. By now adept at quick repairs, I plugged the hole with plywood. I crawled over to the other window and gingerly pushed against it. The entire window gave a quarter inch or so. I pushed it out and away from the roof, plugging the space quickly with scrap.

The weathermen were predicting that the hurricane, now once again in motion, would reach landfall in the early evening. Brownsville was still in its direct path. The television chatter consisted of the meteorologists' relatively unintelligible jargon interpreted, with little expertise, by the local weathermen. Every few hours the Hurricane Center in Miami would issue its pronouncements. The locals began saying not to panic, not to panic. Wind gusts were up to seventy-five miles per hour.

Our electricity went off. We expected it. Often it goes off for no apparent reason. The house became instantly dark. We lit candles, starting carrying around our flashlights, and for dinner ate cold cuts and fruit around the candlelit dinner table.

We began getting the long-distance calls after dinner. It broke up the waiting. First from an old friend in Los Angeles whom we hadn't seen in four years. Then from my father, in Oklahoma. Then my brother in Austin. My wife's parents called from Harlingen, thirty miles up the road. We had no electricity, but the phone lines allowed us to talk to friends and family all over the country. They had heard about the storm. The national news was playing it up big, and as a result, anyone with friends and family in South Texas was becoming concerned.

The winds continued to intensify. I used the portable radio to keep in touch with the reports. The messages became grim. Hurricane Allen was headed directly towards Brownsville. Winds were still predicted at up to 150 miles per hour or more, tides were expected to inundate South Padre and parts of Port Isabel, and torrential rains promised serious flooding throughout South Texas.

I sneaked outside the house to look at the growing storm. The rain had lightened up at least temporarily, and in the dark I could make out the ebony trees and one sour orange in our backyard. The winds were blowing the tops of the ebony trees back and forth as if they were palm trees in the film clips I had seen of hurricanes on television. I sneaked back inside.

By nine o'clock the radio was telling us what to do if the roof blew off our house. In theory we were supposed to find the strongest interior walls and huddle beside them. We were supposed to stay away from any windows and fill our bathtubs. Even the radio announcer did not seem too sure of the wisdom of staying in a house with no roof on it in the middle of a hurricane. A shrillness had entered his voice. I wondered about his family and how he felt having to be away from them.

We dragged pillows and two mattresses into the hallway between the bedrooms. We stocked the hallway with all our necessities. Our teenage daughter read her mystery, occasionally looking up to listen to the radio broadcast. She steadfastly claimed to be more scared of the thriller than the storm. I tried reading but couldn't concentrate. My inner ear was tuned to the wind.

On the eleven o'clock weather report the announcer calmly stated that Hurricane Allen would hit Brownsville in less than a few hours, that the tidal surge from the storm would flow up the mouth of the Rio Grande to flood the downtown and surrounding areas of the city, and that winds would hit us in excess of 150 miles per hour.

My wife and I sat there on the mattress on the hall floor of our house staring at each other in the candlelight. The immensity of the storm, the frailty of our situation, lay between us unspoken. I could picture as well as she a fifteen-foot storm surge roaring up the river, overcoming the levees, flooding the heavily populated Mexican-American neighborhoods. If eighty-mile-per-hour wind gusts could bend ebony trees, wiry and tough as leather, then the effects of winds almost twice as strong, especially on Brownsville's barrios, would be devastating.

I remembered my fifth grade English teacher's telling of "The Birds," the short story that Hitchcock later filmed. The short story, as read to our class every Friday afternoon over a period of months, had emblazoned itself in my imagination. The birds had gone crazy for no apparent reason, attacking the local residents, but unlike the birds in the film, they had been unrelenting. In the film the man and woman escape, but in the story the ending is ambiguous. The two are trapped in the house, the birds beating against the boarded windows with wings and beaks.

The rest of the hurricane night, spent waiting, was like the story for me, only real. I drowsed on and off, fully expecting to hear the roar of the hurricane, like the sound of thousands of wings against wood, ripping the roof to shreds.

But the roar never came, the integrity of the interior walls of our house was never challenged, and toward five o'clock I awoke fully to find that the hurricane, like the birds in the film, had disappeared. Hurricane Allen had deviated at the last moment and slammed into mesquite and cactus fifty miles north of Brownsville, sparing the heavily populated areas. Later I learned from the newspapers that one radio announcer had, at midnight, given up all hope of survival and signed off the air with a "May God have mercy on us." I was thankful that, at the time he chose to predict the end of Brownsville, we had been tuned to another station.

At sunrise I was up and outside scouting for signs of damage to the house and neighborhood. Some electrical lines were down at the end of the block, a few large tree limbs and smaller trees lay scattered on the ground, but no serious damage seemed to have been done in the last two days. The rain had been reduced to a trickle, and a few neighbors were already out clearing away the leaves and fallen branches and stuffing them into green garbage bags.

The electricity remained off for three more days, but we didn't mind. Strange as it may seem, we all felt a definite letdown, a depression of sorts, a sense of having waited for an immense tragedy that never occurred. In a way we were actually disappointed. Nature had showed herself to be supreme, reminded us that she was still firmly in control of our destinies, only to relent at the last hour.

It was not that we weren't glad the city had been spared. We were, enormously. But in us and, I assume, in many others, the waiting had aroused fear and, in the end, panic. I wanted someone, someone in a position of respect and authority, to publicly announce that life could once again return to normal, that there was nothing to be afraid of, that the birds had simply disappeared. The radio, upon which we had come to depend, gave us civil-defense captains, county road-crew foremen, school superintendents, and others who talked in monotones.

The neighborhood, and all of Brownsville, began deboarding. What took eight hours to construct I ripped apart in two and a half. For another three weeks, however, I left up the artificial construction protecting the rear of my house as a totem of Hurricane Allen.

Now, some five years later, Hurricane Allen is a story we tell our new neighbors. They look at us, bemused. After all, the storm did not even strike Brownsville, so why all the fuss? To those who have recently moved to the Texas Coast, Hurricane Allen is a legend, like those other hurricanes that preceded it.

It is just a matter of time, a roll of the dice, before a hurricane, whatever its size, scores a direct hit on a coastal Texas city. Hurricane Alice was just a small sample; its winds played havoc with the skyscrapers in downtown Houston. There are thousands of Texans along the coast who think that a hurricane is just rain and wind with the possibility of a tornado. It is not.

Next time there is a hurricane, we will leave. I will board up the house, load up the family, including pets, and get the hell out. We will watch the storm on the six o'clock news, holed up in a motel in San Antonio.

Sailing to Port Mansfield in the Blue Water Something

We gathered at the Donut Hole on South Padre. The coffee there is always good along with their hot biscuits and gravy. Sailors and wives, girlfriends, a scattering of children, and one well-wisher, we all huddled over breakfast dishes. It had just turned eight o'clock on a Saturday in April, 1978. The sky was overcast, everyone was sleepy. I was seriously questioning whether to move from my chair, let alone spend the day sailing north.

The planning, of course, had been meticulous. The four of us had talked ad nauseam about logistics, including food and camping gear, because we planned to spend two nights on the beach. It was all very basic. Get in two boats and sail north along the intercoastal ship channel. Once in the channel it is impossible to get lost; it is all of maybe a hundred yards wide and the only channel around. The Laguna Madre through which the channel is dug is, of course, much wider. At its end is Port Mansfield, a small fishing community thirty-five miles north of Port Isabel and the town of South Padre Island. We carried a road map, just in case.

By car from South Padre to Port Mansfield one must follow Highway 100 west, catch 77 north, then finally 186 east on what used to be a narrow two-lane piece of asphalt but recently has been widened into another magnificent Texas highway—the kind that gives pleasure to anyone who really enjoys driving. It is a good seventy-five miles that takes an hour and a half or so traveling at the speed limit.

Our flotilla included a Hobie 16 and a Blue Water Something. I can't remember the exact last name, and Joe has since sold the boat and moved to Dallas. I do remember he said the boat was built in Abilene or Denton or some other unlikely place. It was fifteen feet long, gaff rigged, with lots of polished wood. It rode like a fat bathtub—but an extremely secure bathtub—in the water. The Hobie 16 was your basic trampoline on two pontoons, designed for speed. The sail was bright orange with blue marks. The boat was brand new. I chose to ride in the Blue Water Something.

Each of us had brought our own camping gear. I alone had at least thirty pounds of canned food, a tent, a Coleman stove, and four complete changes of clothes, all strapped to a backpack. I had cut by half the cans and perishables that I originally bought at Fedmart. They included three large tins of stew (McGinty) per day, a string bag of oranges (for scurvy, I suppose), three cans of chili, one two-pound bag of rice, a loaf of bread, five cans of Campbell's chicken-noodle soup, and a few extras. I was determined to eat well, and if by chance we should run aground and be forced to wait for rescue, I wanted enough food to get by on. The others had done the same thing, I learned, although relying more heavily on backpacker food and cookies.

The Blue Water Something was designated the supply boat because it was much drier than the Hobie and much more difficult to turn over. By the time Joe had finished loading her with all the gear and the five two-gallon plastic bottles of water, we rode very low in the water. I figure we had easily two hundred pounds of gear among the four of us.

Joe had called the weather service earlier, listened to the recording, then called the other number listed in the phone book and discussed the forecast with the meteorologist. Winds were ten to fifteen knots out of the southeast, partly cloudy with a 20 percent chance of rain. No pressure zones hiding out in the Gulf, no end-of-the-winter Canadian fronts making a final appearance.

The reasons for sailing to Mansfield were more complex than the logistics. The four of us, Joe, Bill, Don, and I, were college teachers. Every year when spring break came along we got the yen to ramble. The yearning is made more acute by working daily with students who look forward to spring break as the last blowout before the short haul to final exams. The feeling is infectious, and we were infected.

We had talked it over countless times in our offices at the college. Other potential voyagers had come and gone, victims of family vacations or projects around the house. By now we had dreams of Mansfield this trip, Corpus the next, and who knows after that. Maybe Houston or New Orleans.

So we set sail after brief goodbyes. The Hobie skimmed away with Don at the tiller and Bill working the jib, while Joe and I sedately ambled along in a light wind, relieved to be underway. We kept close to South Padre Island and headed up the Laguna Madre in the direction of the ship channel. On a map the channel is a pathway of parallel dashes extending along the coast from Port Brownsville to Florida. As the Laguna Madre narrowed from about six miles to half that distance, the sun began to peek through the scattered clouds. A good sign. We found the channel markers and settled in.

Joe and I talked in bits and pieces. There was not much for me to do, because the boat had only one sail, which Joe managed quite well. After an hour or so, however, he let me take the tiller. I am an average weekend sailor. Not a lot of experience over the years, but I have good instincts. Joe is more the kind who reads books on the subject and devises stratagems before he makes his move. He had one story about forgetting to put the plug in his boat and slowly sinking into the bay. Mine concerns the time I took my wife out for our first sail together. I hit a piling ten feet from shore and did a gymnastic somersault over the rudder. She was rescued by a fleet of power skiffs whose occupants had witnessed the gruesome affair. Joe and I had both paid our sailing dues.

We passed Three Islands, the halfway point, about seventeen miles from where we put in. Three Islands are large banks of silt and sludge dredged up out of the channel. Vegetation has taken hold. The fishermen use them as a stopping-off place to stretch their legs. The tiny islands of bay bottom are dotted with the remains of campfires.

Three Islands are the largest of the spoil banks that dot the west side of the channel and can confuse the eye, especially from my vantage point on the Blue Water Something. A system of bays and inlets lies directly behind the spoil banks. Hurricanes, storms, and high tides constantly rearrange the configurations so that at times the mainland actually fronts the channel. Even the most current maps, however, often show the ship channel as rarely bordering the west bank of the laguna.

To the east side of the channel there are additional spoil banks, though fewer in number, and behind them a broad expanse of bay leading to the narrow strip of Padre Island and, beyond, the Gulf of Mexico. Three miles further on we tugged past the mouth of the Arroyo Colorado as the tip of the Hobie's mast finally disappeared from view.

We estimated they would reach Mansfield two hours sooner than we did, maybe more.

I had seen the Arroyo from Harlingen, where it looks like a big drainage ditch. I used to jog along its banks. From the bay side it looks more impressive. We saw a few fishermen in small motorboats and waved. A ski boat zipped by and set the Blue Water Something to rocking.

Joe and I settled into a routine. We commented on what we saw, Joe moved around the boat tidying up, and for the first time we both took a close look at what there was to see. Miles and miles of salt flats, Texas scrub brush and mesquite, the shallow waters of the Laguna Madre stretching out as far as the eye could see. It was very quiet. The southeasterly wind kept the sail full but not straining. The halyards occasionally clanked against the wooden mast. At eleven o'clock the overhead sun cast few shadows. The small waves in the channel silently lapped the shores. The water was a brown and gray mud color that by late afternoon changed to a steel gray.

We sailed past cabins on stilts. They were various sizes and shapes, some quite simple, others ornate. The basic design was a one-room shack on four posts sunk into the sides of the channel. No one was at home. It was like sailing into and through a ghost town of slave huts high and dry on stilts. They all had names and numbers. You lease the land from the state for a nominal fee, then build at your own expense. Hurricane Allen, two years later, would play havoc with these buildings, reducing most of them to driftwood. But their owners would quickly rebuild.

These cabins exist because the small bays and passes to the Gulf offer some of the best fishing in Texas. The drum, trout, and redfish are famous. The flounder, which sometimes reach "blanket size," are worth the effort. To catch them, however, it must be the right season, and the fisherman must know the right holes. These cabins were fishing camps, each hotly in demand a few weekends out of every year, then virtually ignored.

We tied up at one for lunch. I squinted through the boards covering one of the windows. A few wooden chairs, one black potbellied stove. That was it. There were grafitti on the wall, a crooked little dock, and a catfish lying belly-up against a piling. I dug out fruit and bread and peanut butter, one of the extras I had brought along, and washed it down with bottled water. It tasted very good. We stretched our legs, got back on board, and pushed off.

I looked beyond the spoil banks to the west as we lugged along. There were salt flats, shallow bays, walls of brush and cactus leading right down to the water. The channel would occasionally open into a wider bay, giving us quick glimpses of what lay beyond the spoil banks.

The times we stopped for a rest I looked for tracks. I saw a few. Joe said they were coyotes. The deer, if there were any nearby, stayed out of sight in the cool shade.

I know there used to be big cats that roamed the area. At the Yacht Club Hotel in Port Isabel are photographs taken in the 1920s picturing big-game hunters and guides hanging mountain lions from fence posts. In one picture there is a string of eight of them.

Most of the big cats have long since been hunted out, though the local ranchers still occasionally sight one. A graduate student from Texas A&M has been working in Laguna Atascosa, on the way to Mansfield, tagging ocelots and jaguarundi, both about the size of a large domestic cat. They are the only wild cats left, and there are not many of them. They require thick brushland for cover, and that kind of brushland is becoming rare in Texas. I met the student by chance at the home of a friend in College Station. He talked excitedly but matter-of-factly about monitoring the cats by radio. He had given them names.

These coastal brushlands and flats have been used for ranching and, further inland where irrigation is possible, cotton and sorghum farming. The rainfall is no more than thirty inches a year, which technically classifies the area as semidesert. I can imagine what it must feel like to ride through that semidesert chasing down a runaway steer and come smack up against a ship channel. Or at night to hear the howl of a coyote interrupted by the horn of a barge headed for Houston. The land is like this all the way to Corpus Christi, making it a 150-mile chunk of extra dry ranch land, occasionally punctuated by shallow bays, like the Laguna Atascosa, and grasslands. And oil wells. A good part of it is the King Ranch.

I did not see much wildlife. But it is there, at night: the coyotes, the deer, the javelinas, the small mammals, the snakes, the night birds, including the turkey vultures that keep Highway 77 clean as a whistle. On the bottoms of the channel and the bays, the shrimp play in season. The game fish feed off the shrimp and grow to size. The best fishing is often at night from a boat or from a pier with a big overhead light that attracts the shrimp.

By two in the afternoon Joe and I had run out of conversation. Even little bits of talk. The channel was still the same piece of linear water, although its colors had roughened. The shorelines looked flat and dead. I readjusted my sunglasses and headband and pulled up my shirt collar to protect my neck from the sun.

Joe got out his binoculars and scanned the horizon. He could see the tip of the Hobie's mast in the distance. I looked, too. In half an hour we could see Don and Bill and one other guy on a small island only a few hundred yards from the mouth of Port Mansfield's tiny harbor.

The channel opened quickly into a wide expanse of bay. Port Mansfield is on the mainland, South Padre Island eight miles distant. The bay between looks like a very large lake, larger than Lake Texoma. By now we could see the outlines of the water tower at Mansfield, but Padre Island was too far away, too low in the water, to be visible.

The Blue Water Something rocked in the larger swells of the bay. The wind picked up five knots. Bill and Don were still waving at us like crazy. The third guy, in yellow foul-weather gear, didn't move much. Joe and I got worried.

In another half-hour we were within shouting distance, although with the winds at our backs they could hear us long before we them. The first word I heard was "mud." Joe tacked immediately, but it was too late. The Blue Water Something smoothly struck bottom. I got out to push us off and slipped into fine silt up to my waist.

It felt like sinking into piles of smooth silk. In the few seconds it took I remember wondering if my feet were ever going to touch hard bottom. I pulled myself up and out of the muck and onto the starboard side of the boat. The mud in the sunlight was rich and black with the pungent, fertile smell of stockyards and oil fields.

Somehow Joe and I slugged it out with the mud and got the Blue Water Something turned back into the wind. We sailed off another fifty yards and came around on the leeside of the mud island. Don yelled that they were stuck and could not get the boat turned around or lower the sail or any other strategy which would unstick them. Bill, like me a few minutes before, had tried to wade to the island but had gone in up to his shoulders. The mud had sucked away his yellow foul-weather gear, which it now held prisoner a few yards from the Hobie.

I think Bill and Don had a plan for us to free the Hobie, but I never heard what it was. Anyhow, we could not reach them, not even with a line, and they could not reach us, short of carefully swimming out over the mud. That, of course, would still leave Don's new Hobie stuck in the bay with night approaching.

In the meantime the sky had clouded over and the wind picked up. The chop began pushing us around. That was when the home-made skiff loaded with four big men motored by, the only boat we had seen since passing the Arroyo Colorado. We waved frantically. They waved back and approached. They took pity on us. They threw a line, composed of pieces we had on board and what they could scrape to-gether, to the Hobie, and Don attached it to the hull. Then the man at the helm gunned his tiny engine. The motor smoked, the water churned, the skiff moved an inch. He gave it more gas. At one point I thought the stern of his small boat was simply going to be ripped apart trying to move the immovable Hobie.

Ten seconds later Don and Bill gave a shout, the Hobie popped out of the mud with a loud smacking sound, and they scrambled on board. We yelled sighs of relief, and the captain of the skiff pulled an-other beer out of his cooler. Out of an immense gratitude we all wanted to give him something. Don finally gave him a wet five-dollar bill, money to get more beer for himself and his friends. The boatload of fishermen chugged happily away.

Joe turned the Blue Water Something around and headed into the shelter of the harbor. Don and Bill soon followed in the Hobie. We found the public ramp and tied up. Don and Bill looked terrible. They were completely covered with the bay mud from the island, which we later learned was a spoil bank like the Three Islands we had left behind near South Padre. Dried, the mud left an incredibly fine, dark-brown powder. It was difficult to clean and smelled like the inside bottom of those large metal garbage dumpsters behind the department stores.

Later, around drinks and dinner at the Windsurfer, overlooking the harbor, we talked about the day. Our mud predicament had been less frightening than frustrating. The mud was intractable. The showers, beer, and food eased the frustrations, and by ten o'clock our spirits had returned. We laughed the mud away.

The next morning, after Don and Bill had cleaned up the Hobie, we sailed the eight miles over to South Padre Island. A pass cuts the island here into two parts, each accessible only by boat or four-wheel drive. From the north, it is a good hundred miles from Corpus Christi along the beach to the jetties. From the south it is a thirty-mile trip. Consequently the beach usually is deserted and, at least when we were there, free of beer cans and garbage.

About sundown a truck showed up. It made figure eights in the sand, and then the driver and friends pitched a tent several hundred yards down the beach from us and were silent the rest of the night. The isolation we felt, even with strangers nearby, was unique. The Gulf winds and surf were calm. The stars shone extra bright. There were no sounds but those we made.

I cooked my chili, and we sat around our fire in the dunes. By nine, totally exhausted, we fell asleep in our tents. The next morning, after exploring the beach, we headed back to Mansfield. We sailed past several boats fishing for drum in the pass. At the public dock we hauled our boats out of the water, loaded them on trailers, and trucked them back to Brownsville.

As I think back, I realize how little I knew then about the Texas Coast. I had combed the beaches and, on occasion, fished and hunted the bays and wetlands, but my experience outside the places where most people go was very limited. Away from the suntan oil and the beach houses, animals and their sounds predominate along with the wind. I found, once again, that this land can be unexpectedly dangerous. In many of its parts it is in fact a wilderness. The wind, the sun, the water—each is an element that demands our immediate attention and, at the same time, ignores us individually and as a species. Yet the danger, hurricane or muck, can be avoided with a little caution and experience, both of which I lacked.

When we set off for Port Mansfield I thought I was headed for an overnight picnic on the beach. In the thirty-five miles between South Padre Island and our destination I found few signs of twentieth-century civilization. But I did enjoy what at first had appeared boring, then suddenly alarming. There are no Rocky Mountains, Grand Canyons, or other natural phenomena that overstimulate the eye and the heart. There is mile after mile of brushland, thick mesquite bordering a thin strip of sand and a huge expanse of salt water.

Over the months that followed, this trip took on a life of its own as it was retold to others. In my own mind the memory of the spoil bank, and our adventures on and in it, began to recede. What haunted me in its place was the brief glimpse of what I had seen and experienced. As I became more immersed in teaching, raising a family, the day-to-day tasks that encapsulate us all, I realized I needed, had to have, an escape. So I returned to the coast, each time seeing a little more of it, feeling a little more about it, than the time before. The coastal lands became a refuge, a place to hide out, a place to heal under the sun's rays, buffeted by the push and pull of the winds.

Corpus Christi Bay by Butler

Several weeks before I was to drive to Corpus Christi to evaluate a health-care agency I received an invitation from their board of directors. On thick, creamy bond paper was printed in elaborate script the name of the yacht and a request, inside, that I join the board for cocktails, dinner, and an evening tour of the bay.

Usually evaluations of this kind of health-care provider require long days at the clinics and meetings at night with the other evaluators to discuss the findings. A tour of the bay sounded like a pleasant break in the routine and, at the same time, a lot of fun. I have a small sailboat, so I assumed that whoever owned the yacht would be an above-average sailor. How else could he navigate in the dark? I looked forward to learning some sailing from the as-yet-unknown captain as well as finding out more about Corpus Christi.

I chercked into the hotel on North Shore Drive one Thursday afternoon after the three-hour drive up from Brownsville. I got in touch with the head of the evaluation team who had just flown in from New York. She told me that the director of the clinic would pick us up at 6:15 P.M. sharp. I showered, changed clothes, and stood on the hotel balcony, seven flights up, taking a long look at the T-heads below. From that height North Shore Drive and the T-shaped piers which front it were foreshortened; the bay and the dark blue Gulf sparkled beyond.

I was glad the winds were light and the chop on the bay at a foot or less, according to the marine report. I did not look forward to getting seasick, which I was sure would happen in someone else's boat. (As long as I am at the tiller I seem to have no serious problems. But let someone else do the steering, even on my own boat, and my stomach begins rumbling and I am liable to become disoriented.)

The director of the agency met us in the lobby. She was intelligent, vivacious, and (as I would be in the same situation) a little nervous. In the next two days our evaluation team not only would give her

bosses our analysis of the job she was doing, but would also take a serious look at all aspects of the health-care facility for which she was
responsible.

I was surprised by the director's appearance. She wore an evening
dress complete with fur. She mentioned on the short ride over to the
T-heads that this would be her first trip on the boat, owned by one of
the board members, but that she had heard great things about it.

After a few false starts we found the boat. Actually it was not a
boat: that word did not apply. It was a ship, berthed diagonally at the
end of one of the T-heads because it was so large it could not fit any-
where else. At first we simply did not see it; it was too big to notice. I
had my eye on some of the thirty- to thirty-five-foot motorsailers and
completely ignored the hulk at the end of the pier.

I would guess the ship at at least 110 feet in length. She had three
decks and, compared to the boats nearby, looked like a floating castle.
There was a little gate next to the gangplank we strode up. One of our
hosts greeted us warmly and pointed the way to the "salon."

I followed the two ladies and the hostess into the salon. One step
in and I felt I had entered a time warp. I was on a 1920s movie set. The
butler asked for and took the director's fur. He was dressed in tails and
had a European accent. The maid scooted out a side door. The bar, to
my immediate left, sparkled in chrome, mirrors, and finely polished
woods. Off to my right a group of men and women sedately stood with
drinks in hand.

Trying to remain blasé about the spectacle, I ordered up a drink
and sat in one of the antique leather chairs near the sofa. The salon was
incredibly well preserved. The wainscotting gleamed as if it had just
received a coat of polish that very morning. The brass, the chandeliers,
every detail seemed to have been precisely groomed and burnished.

Once having realized that my own boating standards of Gatorade,
Coors, and hamburgers were definitely out of fashion in this group, I
regained my composure and began to mingle. The small crowd was the
usual mix of doctors, lawyers, and health-care administrators. Sprinkled among them, however, were two oil magnates, a grand dame, a
newspaper publisher, and a few dilettantes.

Appetizers were served. The chef appeared, promising (in a
French accent) great things from his stove. Before he left for the galley
he received abundant praise from those who knew his work. While the

chatting continued, the ship majestically, almost imperceptibly, pulled away from the pier and slowly dieseled out toward the bay.

The evening proceeded as planned. The guests were given a tour of the ship, which, the hostess assured us, had been owned by Al Capone. She took us down some stairs to the first deck, where we were shown five staterooms, four bathrooms, and a huge vestibule. The first deck had been recently remodeled. The staterooms reminded me of your average expensive condo. Some of the original bathroom fixtures, however, were still in place, including the ornate patterns in tile that were popular when the ship was constructed.

The hostess proudly showed us Al Capone's bedroom. I read about his life, his rise and fall, in old newspaper clippings that hung in the hallway outside the stateroom.

The bell was rung for dinner. The company was divided into two groups, those who sat around the large table in the captain's dining room, and those, presumably less important, who remained in the salon behind little wooden folding tables that the butler brought. The captain's table was a magnificent piece of furniture. The rich woods obviously had weathered the last sixty years well but did show the nicks and scrapes of time that gave them genuine character.

The food was excellent. Buffet tables had been set up on the aft deck, accessible through a door from the salon. The deck outside was partially protected by a canvass awning. I made every attempt to convince the chef that I respected his work by devouring as much as humanly possible. I chose roast beef and rarer delicacies, at the same time staring into the wake of the ship's powerful engines. My plate piled high, I returned to the dining room.

The conversation over dinner was light but intriguing. Toward the end of the meal one guest, who'd had too much of the excellent wine, began sermonizing. The hosts duly ignored him, directing the conversation into other channels. The butler came and went. I discovered that it was fun to tell the butler what I wanted. He left, returned, left again. I enjoyed being catered to.

After dessert, chocolate torts with coffee and brandy; the only thing missing was a twenty-dollar cigar. I realized that being very, very rich certainly had its attractions.

I waddled out through the salon door and stood in the Corpus Christi Bay fog. During dinner I had completely forgotten we were touring the bay. The ship flattened the small chop, and the only hint of real movement was the muted drone of the diesels. From the bay side, Corpus periodically sparkled like a jewel between puffs of ground fog. The publisher joined me, and we talked briefly before going back into the salon.

That night I had little interest in or appreciation for the city with its twinkling lights fronted by the black bay waters patched in fog. Nor did I see the pilot house. The machinations of the ship's human cargo were far more fascinating. Toward eleven we returned to the T-heads, and the next two days we spent in serious study of the health facility.

There are many ways to enjoy the Texas coast, its beaches, and bay waters. One way, if you have the money, is to take your mansion to sea.

The Biggest Oil Spill in the World

In early June, 1979, Ixtoc I sprang a leak in the Bay of Campeche. The drilling platform was located in the southern quadrant of the Gulf of Mexico, hundreds of miles from Texas beaches. It was leased by Pemex, the Mexican oil monopoly, from Sedco, a company founded by then Texas Governor Bill Clements. As the oil spilled into the waters surrounding the rig, the seasonal winds began pushing the sludge northward, towards Texas.

It is estimated that Ixtoc I dumped more than 150 million gallons of crude oil into Gulf waters. This ranks the spill as the largest in the world, almost three times that of the previous record which resulted from the break-up of the *Amoco Cadiz* off French shores in 1978. While the immensity of the oil spill is impressive, even more so is the fact that so little was ultimately learned from the biggest oil spill in the world.

In the early summer months of 1979, residents of South Padre Island paid scant attention to the spill. A few local environmentalists were worried about the marine wildlife and beaches near the rig, but their concern was shared by few others. As the sludge began its slow trek north, batteries of scientists in the local media assured the public that there was nothing to fear. They reasoned that the distances involved, not to mention the effects of wind, weather, and tides, would keep the crude far south of the Rio Grande.

Through mid-July no one paid much attention to Ixtoc. In the meantime, Pemex officials used various methods in an attempt to cap the well, which nevertheless continued to spew oil unabated into the Gulf. The front line of oil, disregarding all the elements that were supposed to destroy it, continued to move northward.

In late July the scientific experts began to shift their predictions. They accomplished this by sleight of hand: their denials that Ixtoc oil would reach the United States became probability statements, percentages of likelihood that the spill would cross the border. And in the following weeks these probabilities changed drastically, ultimately expressing confidence that the oil would reach Texas shores. People all along the coast began talking about contingency plans.

Few residents of the Texas Coast were aware of the coordinated efforts of the United States Coast Guard and the National Oceanic and Atmospheric Administration, aided by the Department of the Interior, the Department of Defense, and the Environmental Protection Agency. Corpus Christi was chosen as the major base of operations for the coordination of the defense against Ixtoc. Computers were set up, observation flights of the massive slicks begun, and plans drawn to react to the spill should it threaten the Texas Coast.

Rumors abounded. The newspapers talked of sealing off the Brazos-Santiago Pass to the Laguna Madre to keep the oil from spoiling the spawning grounds of shrimp and fish. Further north, residents in Aransas Pass grew concerned over the fate of the endangered whooping cranes if they were contaminated by the Ixtoc oil. Rumors spread.

As the ooze moved northward, time passed slowly, a slow waltz that caressed the fears of those who lived along the coast. The tides came and went. There was absolutely no sign of the goo on Texas beaches. The experts continued their reassurances (although their bets against the oil soon would reach an even fifty-fifty chance).

Many local residents believed the oil would never come, that it could not, that their beaches were somehow immune. The majority of experts, their current denials to the contrary, upheld the public's confidence that some kind of miracle would take place, that the crud would magically disappear. Only when plane sightings of twenty-mile-long oil slicks became everyday fare on national news did the reality of the event impress itself on those who lived along Texas shores. The experts began to disappear into the woodwork as the sludge approached the Rio Grande.

The slimy goo hit South Padre Island in the first week of August. The beaches were closed so that work crews, laboring under the steaming summer sun, could do their job unhindered by crowds of sightseers. The Coast Guard threw up a floating barrier across the Brazos-Santiago Pass to deny the oil entrance to the fragile ecological system of the Laguna Madre. Farther up the coast, similar maneuvers were attempted. The tourists stayed away in droves. The normally thriving town of South Padre Island was empty this summer, only locals and residents from surrounding towns bothering to visit the beaches for a peek at the oil. One view was quite enough; most people I talked with said they would return only when the beaches were totally free of the sludge. The crud moved northward, up to Port Mansfield, across the long boundaries of the King Ranch and the smaller ranches that border the bays and inlets of South Texas.

Ixtoc continued to leak its thousands of barrels of crude. The work crews along the coast kept up the cleaning operation, using shovels, straw, and sweat. The subcontractors for the Coast Guard ran hither and yon claiming local victories for the environment only to face another inundation of Ixtoc oil after the next high tide. Finally, in mid-September, after six weeks of a losing battle with Ixtoc, the winds changed direction, blowing the sludge out to sea. Coastal Texans heaved a collective sigh of relief. Six months later Ixtoc was contained at its source.

Politicians visited the oil-covered beaches and, in spite of what they saw, claimed in the media that the expected disaster had been averted, that tourists could immediately return to enjoy Texas beaches. Much of the local media went along with the charade. Still the tourists stayed away.

Even as the oil was coating the beaches, the political chorus minimized the existence of the oil spill, redefined it as a nondisaster, an ecological calamity that never happened. The national media soon lost interest, as is their way, and the local television stations and newspapers followed suit. That left only the legal side of the nonproblem, soothing those who lost tourist dollars and paying for the cleanup of the nonspill.

Sedco, owners of Ixtoc, reached a settlement with claimants to the tune of two million dollars. Most officials of the coastal communities polluted by the nonspill believed that, at best, the settlement would cover only a portion of the real losses. Sedco, meanwhile, said it planned to sue the Mexican government for the money it lost because of the lawsuits.

Most of the researchers and experts who had called the shots before Ixtoc oil reached Texas beaches simply vanished. Since there was, by redefinition, no spill, there was no reason to study seriously the effects of the oil on Texas shores, bays, and wetlands and in the waters of the Gulf of Mexico. There was no flood of federal or local monies from state universities or foundations to fuel research on the Ixtoc mess. Without the dollars, the scientists who would have studied the largest oil spill in the world took their research interests elsewhere. (Unfortunately, too, Ixtoc itself was never examined by independent observers to ascertain the exact causes of the spill. The rig was hauled out to deep water and scuttled.)

A few studies have appeared in the research literature. However, by the fall of 1985, more than five years after the oil hit the beaches, no major research effort offered a broad, empirically based analysis of the impact of Ixtoc on Texas or Mexico.

The absence of major research is rationalized away by the observation that the greatest damage had been done in Mexican waters and on Mexican beaches. If the Mexican government wanted the research done, so the thinking goes, they were the logical ones to sponsor it. That research, of course, has never taken place; Mexico's grave economic problems take precedence over its concern for protecting its environment. Thus, we lost an excellent chance to observe the effects of a major spill on marine wildlife, coastal ecosystems, and those who live from the sea.

That fall the phone calls started. For the next two years, approximately every four to six months, two or three men working as consultants for major oil companies would call me up. At the time, I was going up and down the coast interviewing shrimpers and others in the fishing industry. The consultants wanted to know, in so many words, if I had heard any shrimpers complaining of poor catches as a result of the oil spill. Barring that, they wanted word of any work-in-progress that could demonstrate a relationship between decreased shrimp production in the Gulf and the oil spill. I knew (and know) of no such projects and told them so, repeatedly.

The major oil companies, unlike any governmental agency, were concerned that the spill had brought about some significant biological changes in the Gulf. They worried not only about the threat of lawsuits resulting from such damage, but also about the publicity that such suits inevitably would bring. Big oil has prided itself on being a "clean" industry, relatively speaking. This image means significant support in Congress when it comes time to pass laws sympathetic to the majors. It was in their best interest that Ixtoc remain an insignificant blemish, a foreign aberration, on their environmental record.

What really saved the Texas beaches from Ixtoc crude, it needs to be emphasized, was a change in seasonal winds, a change that came unexpectedly early in the year. The intergovernmental effort was an improvement over earlier responses to spills, but what is truly impressive is that in spite of the agencies' efforts, so little could really be accomplished. Beneath the maze of computer simulation models, flights employing sophisticated electronic gear, and twenty-four-hour hotlines on the status of the spill lies a cold, hard fact: there have been no technological breakthroughs that would expedite the containment and clean up of a major spill.

Said another way, if the winds had not changed when they did, all the government's scientists and their support people would have been knee deep in crude.

The phone calls from the consultants stopped six months before the final settlement between Sedco and the claimants. Unfortunately, the Ixtoc I spill, along with previous contaminations of the Gulf of Mexico, has been relegated to history.

In July, 1984, the tanker *Alvenus* dumped 66,000 barrels of crude oil off the Louisiana coast. Even though the oil spread over Galveston beaches, it caused no long-term damages, but once again it served to remind us of the vulnerability of the Texas coastline. We will, unfortunately, have to face the impact of major spills again and again. Only by studying the causes and effects of spills, prevention techniques, and methods for collecting damages can we ultimately hope to preserve the Texas Coast and the Gulf of Mexico from man-made disasters. We got lucky with Ixtoc; the winds were in our favor. But our luck cannot hold forever.

Castles Beyond High Tide

Last Sunday my family went to the beach. I plopped down in the sand just beyond the reach of the waves. Travis sat down beside me. We were armed with two shovels and two pails. Andrea strolled up the beach with Lauren while Travis and I got down to the serious business of building a sand castle, one that I was determined would last forever.

He took the large blue shovel and began stabbing at the packed sand. It broke into small chunks, which he laboriously shoveled one by one into the yellow plastic pail. I was left with the smaller shovel and the white rusty pail, still filled with remnants of the last trip to the beach. I dug up some sand quickly, molded it into a mound, then told him that the castle was beginning. He immediately defined my small mound of limp sand as my "house" and his still smaller mound as his "house." I said it was a castle; he said it was a house. I kept building my castle; he worked with intensity on his house.

For ten minutes we built. Around us the wind blew gently, the sea gulls, unmolested by the usual summer crowds, drew closer, and the waves continued arriving in one inundation after another, an infinite line of energy reaching into the Gulf and beyond. My son ignored all but his labor. Sand appeared on his knees, arms, and face. Oblivious, he worked on under the noonday sun.

I excavated a moat around the castle, keeping a wary eye on each new wave that reached near us. The moat would save the castle, siphoning off the Gulf's forces. A wave did finally sweep up to the new castle, in part buffered by the moat, then receded. Travis looked shocked, then smiled as the wave disappeared and the castle, with newly rounded corners shaped by the salt water, still stood. The moat was temporarily filled. He splashed in the six-inch water then turned with new vigor to the building of his house.

I drained the moat by digging a tiny channel back down the slope of the beach to the surf. I finished the moat, topped off my turret by dripping wet sand from my hand onto the mound, now a foot high, and shifted in the sand to help Travis. Sand had worked its way into odd places on my body, between my toes, behind my ears, and up the sides of my suit. If I think about all the places it seems to seek on its own I get irritated. Ultimately, of course, it somehow always finds its way into my mouth, where, grit between my teeth, it reminds me of the dentist's chair, drilling equipment, and spitting into the small lavatory with its tube of running water.

My son worked on, consistently but inaccurately, slowly creating his house in sand. The noon sun cast a bright, flat light on his thin shoulders and upper torso; the skin on his shoulder blades began to turn a light, delicate pink. The breeze ruffled his curly blond hair. The occasional walkers-by would stop, stare at the two of us, father and son, then walk on down the beach.

Taking a break from our labors, we waded into the waves, knee deep. I held him by the wrists as the power of the waves pushed against his small body. At one point he fell, clinging to my ankle with both hands as the force of the waves pushed him back toward the dry shore ten feet away. He laughed and giggled with the fun of it, the wonder of it, the enjoyment of waves, sun, gulls, and sand. A small gray wave slapped against his face, cleaning it, temporarily, of sand. He swallowed some salt water, spit it out, made a face, and grabbed once again for my hands.

I picked him up, my arm around his waist, and lugged him like a sack of potatoes back up the slope of the beach. The castle stood as before, but now the moat was full again, the drainage ditch washed away in the last set of waves. Travis sat down in the moat, his attention riveted again on his house. Andrea and Lauren returned from their walk. My daughter crawled around on the castle mounds, excited by the unfamiliar beach sounds and sights. At nine months of age she already is demanding her share in what we do. She will not sit and watch; she joins in, participates at her limited level, and seems satisfied.

Again I drained the moat. I reinforced the castle walls, making them twice their original thickness. For good measure I built a smaller structure immediately behind the castle. Even if the castle fell, the tiny fortress might still survive.

In another five minutes the three of us were covered in sand again. I rinsed Travis in the waves. Andrea trudged back across the beach to where the car was parked. I grabbed Travis in one arm, buckets and shovels in the other. He did not resist. We went through our litany of goodbyes to the beach, including one goodbye to the sand castle we both had built.

From the parking lot the sand castles were barely visible. In another hour they might be gone, washed flat by the encroaching waves. We, my wife and I, dusted off the sand from our children and ourselves as best we could. On the way back home, Lauren, ensconced in her car seat, quickly went to sleep, tired to the bone by her efforts and adventures.

Travis sat between us on the drive home. Forefinger in his mouth, eyes puffy, sand hiding between his curls, he began to nod back and forth in the first throes of what would be a heavy sleep.

Castle sand stuck tenaciously between my toes, on my ankles, and to the seat of my trunks. Sand between the steering wheel and my hands made turning an exercise in grit. Still, it was not an unpleasant sensation. The warmth of the sun on my back and the open spaces of the Gulf combined to create a feeling in me of timelessness and closeness to family, to those I love.

I leaned over and told Travis that next time we would build a new sand castle, a bigger one. He looked up at me and nodded yes. Change, new waves against old castles, he accepted without question. He did not build his castle, his house, to last forever; he built it to serve his momentary, intense purpose. I, on the other hand, had thought from the beginning of nothing but how long the castle would last and what could be done to prolong its existence.

Later, in the shower, I washed the last grains of sand from my son and daughter and myself. As I finished washing each child, Andrea caught them in a towel and carried them off to their afternoon naps. I finished my own shower, shoving water against the last few grains that clung to the enamel. The sand went down the drain and out the line into the sewers below the streets, where eventually it would flow into the Rio Grande River and, once again, rejoin the waters of the Gulf of Mexico.

Toweled off, cozily dry, I sat on the sofa in our large back room, which serves as nursery and den. Andrea asked me if I wanted to take a nap. I resisted. There was a football game just starting the third quarter. It was a close game between two good teams. I couldn't pass it up. She asked me again if I didn't want to take a nap. I thought about a nap, the sun at the beach, our sand castles now under high tide. I slept the same peaceful sleep of my son and daughter.

Cannibals

Although raised in Oklahoma, I never met or read about American Indians until I left the state. Oklahoma has one of the largest Indian populations in the country. I remember seeing them walking along the street, especially when my mother took us to her home town of Bristow, not far from Tulsa. I do remember the Indian in full battle gear at the annual Oklahoma-Texas football game. He would go into a frenzy every time OU scored a touchdown. The band beat on their drums as he performed intricate dances, strutting his feathers and paint.

When I heard about the cannibals along the Texas coast, fierce Indians who consumed their captives, I was skeptical. I reasoned that coastal Texans, like their neighboring Oklahomans to the north, probably knew little about either the history or the culture of these particular Indians. Everywhere I went along the coast, however, I heard the same stories about one tribe in particular, stories that had a remarkable consistency. I came across references to the Karankawas, for so they are named, in magazine articles, local histories, and Chamber of Commerce literature.

Three things were always part of any story I heard. First, the Karankawas were always described as being tall, the men approaching seven feet in height. Their inhuman smell was always compared to that of various animals. Finally, the description of their voracious appetite for human flesh, whether that of Spanish explorer, European settler, wayward priest, or enemy Indian, was always related with a certain morbid gusto.

I did some library research because, since leaving my home state, I had grown to distrust what we often know of our own history. I found that, indeed, the Karankawas were tall, especially when compared to the first Spanish explorers, who sometimes did not clear five feet. The male Karankawas often reached six feet in height or taller, only a little above today's average but much taller than the average European in the sixteenth century.

The historians likewise document the odor of those Indians who first came into contact with whites. They must have smelled awful, coated with some combination of animal grease and who-knows-what. Of course, the more I thought about it, the smarter those Indians sounded. After a heavy rain the mosquitos and biting flies can overwhelm an innocent visitor to the wetlands or beach. I've run through clouds of mosquitos and have watched from a safe distance as others, when surrounded, have shouted and yelped and cursed those insects. They get into your eyes, ears, and nose, any available space, and they bite through shirts to the skin beneath. There have been times when I would have welcomed a coating of alligator grease insect repellant.

I also found descriptions of the Indians' weird daily ritual of taking baths in the bay waters. The European explorers bathed irregularly at best. They believed the Karankawas to be idiots for washing themselves on a regular basis. The Europeans presumably piled layer after layer of perfume upon their bodies and became accustomed to each other's odor. In fact, the smells that collected on the Europeans' bodies over the long winter months probably served as effectively as alligator grease to drive away the mosquitos and other insects. Again, there is a bias reflected in the present-day folklore that the Indians smelled bad. I wonder what the Indians must have thought of the white men who seldom bathed?

The third characteristic, that the coastal Indians were eaters of human flesh, is a much more serious charge. The historical evidence is quite strong; the Karankawas were not cannibals. The allegation has been refuted both by the Texas naturalist Roy Bedichek in his *Karankaway Country* and by the anthropologist W. W. Newcomb, Jr., in his *The Indians of Texas*. The truth has been ignored not just by magazine writers and tourist bureaus, but by many Texas historians as well.

Withoug belaboring the point, accusations that the Karankawas ate human flesh are always third-, fourth-, or fifth-hand, never eyewitness accounts. Not a one. Cabeza de Vaca, who lived among the Karankawas for some years, never mentioned cannibalism in his journals. The practice of cannibalism is not found in neighboring coastal tribes to the north or south. To date, no archeological evidence suggests cannibalism occurred among these Indians. The only well-documented account of cannibalism, in fact, is that of Spanish explorers who, when shipwrecked on a barrier island, were reduced to eating their dead comrades in order to survive.

Both Bedichek and Newcomb offer a solution to the cannibalism puzzle: by branding these Indians cannibals, Europeans who explored and finally settled the Texas Coast justified their policy of extermination. By 1827 the Karankawas had been reduced from an estimated population of ten thousand to a pitiful band of no more than one hundred, who, forced to sign a treaty with Mexico, were heard of no more.

This does not mean that the Karankawas were innocents. The historical accounts of Karankawas stealing from, kidnaping, and murdering white settlers are extensive—as are the accounts of similar European acts against the Indians.

What we have here is more than a historical lie, a harmless piece of Texas folklore. The belief in cannibals reflects in part what we hold as true about the coastal lands, namely, that the lands have no real human history. Cannibals are somehow subhuman, not quite members of our species. Thus, according to the lie, human history began along the Texas Coast with the Spanish exploration in the sixteenth century. In truth it began thousands of years before. The lie prevents us from taking an interest in, or learning from, people who knew the coast most intimately.

Over the eons the Karankawas developed a way of living precisely calibrated by the lands upon which they depended. Hunters and gatherers, they neither raised crops nor tended animals. As the seasons changed, as food supplies diminished, they moved on. They carried their homes on their backs. Using poles and branches they built temporary structures that sheltered them from the winds and rains of fall and winter. They had few possessions; those that they carried from one place to the next were absolutely essential to their survival. Over thousands of years they nurtured a knowledge of the animals, the plants, the places upon which their existence depended. The knowledge was assiduously passed on from one generation to the next by word of mouth.

Throughout human history the Texas Coast has been a niggardly and barren land. The Karankawa Indians learned how to fish the bays from their dugout canoes, exploiting the trout, mullet, flounder, shellfish, and other species that still frequent the waters today. They killed deer, lived off small mammals and rodents, and learned which plants and plant parts were edible and nourishing throughout the year. Starvation was always a possibility in the lean times. The total population, which probably never exceeded ten thousand, was strictly limited by what the land could offer.

In the twentieth century we have, with the aid of technology, installed irrigation systems that bring fresh water to the coastal lands. Consequently, farming and ranching have become the major industries in most Texas coastal counties. We have tapped the oil and gas reserves in the bays and estuaries and, using offshore rigs, have tapped the rich oil fields beneath the Gulf of Mexico. But the coastal land itself has not changed. It is still niggardly, a land that, because it borders a sea, is always in transition.

Today we are rapidly building condos and accompanying buildings where no one ever before built permanent structures. We construct them just above the mark of high tide on land, much of it sand, that borders our bays or fronts the Gulf of Mexico. In the 1980s small fishing communities are being replaced by tourist resorts from South Padre Island to Port Arthur.

But neither the essential resources of the coast nor our requirements for survival have changed. First we install bridges between the mainland and barrier islands, islands that by definition are temporary. Next we import fresh water along with food, electricity, and the other requirements for a modern life-style. Then, when there is not enough water to go around, we complain because we cannot water the Saint Augustine grass in our front yards. And we complain even more bitterly when our precarious communication, transportation, energy, and waste systems show inevitable signs of strain.

Without the benefits of modern technology the Karankawas were forced to accept the limitations of the coastal lands. Certainly I have no desire to return to those times. Nevertheless, as I sit here in the latter part of the twentieth century, it seems obvious to me that these indigenous Indians have something to tell us. The Texas Coast is not inexhaustible; its resources are extremely limited. Nor are we particularly wise, for the sake of tourism and recreation, to erect a twenty-story condominium on shifting sand and mud, the Gulf of Mexico on one side, a turbulent bay often on the other.

Bedichek ponders the image of a Karankawa warrior described in a firsthand account by a settler—a warrior who stares into the sun as it disappears into the bay waters. Bedichek wonders about the Indian's belief system, a system that probably ascribed magical powers to the setting sun. We know so little about these Indians. But we do know that they were not cannibals, that they were human beings like you and me.

I cannot help but think that the more we know about the Karankawas the better off we will be. My instincts tell me that there are secrets to be learned by archeologists and anthropologists who explore the Indians' culture and carefully reconstruct their history before the onslaught of the Europeans. The wisdom that would guide us in future uses of the Texas Coast lies, at least in part, in shell middens as yet uncovered. I lie awake some nights, guessing at what the Karankawa warrior must have been thinking as he watched the sun, an inverted mushroom, sink yet again into the salty bay waters.

The Bay Sun

I stood with two newfound acquaintances on the balcony of a restaurant overlooking the bay. As the sun began to set, we looked studiously down at the throngs lining up to record the event on film. The two, grocers from Tennessee, and I poked fun at the tourists who had stopped en masse between margaritas and Singapore slings to gaze at the nightly occurrence. Someone rang a bell, a voice came over the recorded music to the effect that the sun was indeed setting, and the cameras began to click. If our species does not have the time to understand something, we take pictures of it. The rim of the sun's glow disappearing into the bay waters, the temporary silence ended as the tourists focused their attention on each other once again. The recorded music was turned up a notch by an unseen hand.

I'm not exactly sure what makes bay sunsets different. In part it may be the colors and forms, unmuted by haze or smog, blending liquid fire with bay waters. There are only the sun and the water, then the sun melting into the water, then the afterglow that lights both water and darkened sky.

I've seen more impressive sunsets in other places. New Mexico comes to mind. With others I have perched on the mountains that overlook the Rio Grande Gorge at Taos to watch the dramatic exit of the sun as shadow and light play off the colors of the desert. Those times left us in wonder, incredulous that the event could be repeated the next evening.

Sunsets over Texas bays are more prosaic. There are no waterfalls, rivers, or Big Sur redwoods as backdrops. Just an occasional palm or windswept oak like those at Rockport. It is almost always a simple event. As dusk eases into night I am left with a satisfied feeling of one more day having passed by, an ordinary day. I am not held speechless by the sight of a Texas sunset over the bay, but I am more assured in the regular workings of the planets and stars.

In fact the time we call twilight has more magic for me than sunset. It is the time when the bay waters ripple in their diamond sparkles, setting off tiny blasts of light that attract the eye with their pinpoint velocity, then disappear. It is a time when a seabird glides low above the water, headed homeward. It is the time when some of the tourists hesitate between words, stare off to the west, and lose their train of thought to the larger processes that surround us all.

Some firsthand observers noted that the Karankawa Indians stood watching sunsets for hours. My guess is that for them a sunset over the bay was just the beginning of their twilight show, a form of personal entertainment as well as of reassurance. Without need of Singapore slings or cameras they stood observing a light show that faded into the dark shadows of the night. The fade to black, which meant they must return to smoldering campfires and sleep, was quite possibly the high point of each arduous day's struggle with their environment. Perhaps for them it was, too, a proper ending to another day, a day like all of those that would ever come.

When the Condos Come to Town

At least once a week I jump in my car and head for Port Isabel and South Padre Island. If the weather is warm I store the sun roof in the back of my car, roll down the windows, buy a Coke at the local Maverick Mart, and take off. Behind me, in Brownsville, I leave job and family worries. Before me, stretching like a lazy snake in the sun, is Highway 48. Through land the Karankawas once roamed, the road now slithers past welding shops, junkyards, cantinas, Madame Palm's fortune-telling business, construction companies, marine suppliers, and a wholesale beer distributor.

I decompress from life's daily anxieties as I shift my car up through the gears, feel the rush of the wind on my face, and smell the familiar mixture of diesel fumes and salt air near Port Brownsville. To my left is the Union Carbide plant, now deserted, illuminated at night like a city of ghosts by ten thousand lights, not a person in sight. On my right in the ship channel are two freighters, each being demolished for scrap metal, which is loaded onto flat cars and railed to Monterrey and other points south.

Once past the Marathon Le Tourneau plant and the shrimp fleet immediately to the east, I finally reach open country. I let out a sigh of relief. Now there are salt flats and marshes on either side of the highway. During the late spring and summer the surrounding sand, stirred up by thirty- to thirty-five-mile-per-hour gusts, coats the interior of my car, my clothes, everything it can reach, with a finely powdered layer of minute crystalline pellets.

As I round a large bend in the road, the skyline of Port Isabel first becomes visible. I turn right at the intersection of Highways 48 and 100 and slowly motor into and through the outskirts of Port Isabel. This particular stretch of highway is always patrolled by local police armed with radar guns. Many of my friends have fallen prey to one of the best speed traps in Texas. I spot the dark blue sedan parked at its usual spot and slow down to a stately twenty-five, ten miles per hour less than the speed limit.

I pass the football stadium, home of the Fighting Tarpons. Lettering on the press box announces to all that the team has done well over the last thirty years. Last spring a prankster replaced the *r* in Fighting Tarpons with an *m*. Local residents were not pleased; the football team is the pride of the community.

When the weather is particularly dreary, like today, and even a walk along the beach is not possible, I turn left at the Dairy Queen and head for the Yacht Club Hotel and Restaurant. The Spanish-style building, easily the finest piece of architecture in Port Isabel, sits directly across from the small boat harbor that links the town with the Laguna Madre and the Gulf. Perched, finally, on a bar stool, I take yet another look at the enlarged photographs that line the main dining room of the restaurant.

I stare at photographs of another era, of grander times, when the structure in which I sit was first erected and used as a real yacht club by its founder. He built it in the 1920s so that he and his friends could enjoy the boating, fishing, and relaxation that Port Isabel and Padre Island had to offer. There are photographs of the wooden-hulled boats they sailed, the strings of trout they landed, and their favorite swimming holes. There are also pictures of proud hunters next to the carcasses of wild cats which once roamed the area in abundance. To one side stands the local guide; to the other stand the hunters with their rifles.

Back then fish camps on South Padre Island served as bases for trips made northward along the coastline. Hotels, one or two in the grand style, were erected on this barrier island and north of the mouth of the Rio Grande on Boca Chica Beach. None withstood the hurricanes, and eventually, by 1940, the attempts at permanent structures were abandoned.

The fishing camps, composed of a few wooden shacks and other limited facilities, typically endured the storms and could always be rebuilt if necessary. Fishermen reached these camps by chartering a boat with a local guide or, eventually, using the ferry that used to run from the end of old Highway 100, five blocks south of where the new causeway, erected in 1967, crosses the bay.

Port Isabel, first settled in the early 1830s, depended almost exclusively on a small fishing industry and the income it generated from tourists, like those that visited the Yacht Club, for more than 120 years. Men who could not find work at the surrounding ranches eked out a subsistence from trout, redfish, shrimp, oysters, blue crab, clams, flounder, and scallops in the Laguna Madre. By the 1900s, Port Isabel could boast of one fish-processing plant and several packing houses. Entire families worked, when needed, in the plant and the packing houses, sorting, icing, and transporting the fish to market.

On the third floor of the Cameron County courthouse in Brownsville is a small collection of photographs of Port Isabel at the turn of the century, photographs some twenty years older than those at the Yacht Club. The tiny village of Port Isabel clings to a sand spit projecting into the bay. On the west side is an endless stretch of South Texas mesquite and scrub; on the north, east, and south are bay waters. It looks to be, and was, a precarious existence, a life spent in isolation from nearby towns and cities. Highway 48 is a relatively recent link between Port Isabel and Brownsville. It was not so long ago that to get to Brownsville from Port Isabel one was required to travel by way of Harlingen, a distance of approximately sixty-five miles.

Generation after generation of those raised in Port Isabel either left to find work elsewhere or remained to marry within the small community and live out their lives in relative seclusion from the rest of the world. As a result, politics in Port Isabel today is based more on one's name than on any specific qualifications. Several Port Isabel families have been arguing among themselves for generations.

In the last ten years the condominiums have come to Port Isabel. The tiny boat harbor is now bordered on its southern side by a row of three-story condos priced in the $130,000 range. Where Highway 100 now spans the bay with its four lanes, a huge marina is planned on as yet partially submerged land leased from the city. A few blocks to the south of the causeway stand several new developments with condo-lined channels. Across the swing bridge, on land recently filled, sits a large recreational vehicle park complete with pools, tennis courts, clubhouse, and a long row of condominiums.

I order a scotch at the Yacht Club bar and think about both the rapid influx of condos to Port Isabel and the changes yet to come. Only once before has Port Isabel undergone such major changes. Immediately after World War II Louisiana shrimpers relocated their Gulf trawlers to tiny Port Isabel and Port Brownsville. The Cajuns brought with them not only their boats, huge by local standards, but also their knowledge of the world beyond South Texas. The Cajuns spoke a different language, ate different foods, had different hopes and dreams for themselves and their children.

The gulf shrimping industry was an economic blessing for Port Isabel. It provided jobs for those who were barely getting by. The new capital revitalized the community, contributing to a growing city tax base, upon which depended both the school system and basic city services. Eventually the most ambitious men in Port Isabel bought their own Gulf shrimp boats, built themselves finer houses, and became respected businessmen in the community. The community prospered, and many shared in the new prosperity.

For the next thirty years, from the 1940s to the 1970s, the fortune of Port Isabel hung on the ebb and flow of the shrimp industry. And ebb and flow it did from one year to the next, but always, over the long run, the industry grew. While the shrimp one year might be few and far between, a patient shrimper and his family knew that if they held on for a year or two, the good times would return. The fifties and sixties, when surveyed as a whole, were the golden years of Texas shrimping for Port Isabel and other Texas communities that rely on the sea.

Just as the shrimping, besieged by higher fuel prices, stiffer federal and state regulations, and increased competition from imports, began to fade away, the first condos appeared. Local developers, discouraged by the high price of land on South Padre Island, naturally came looking for a better place to build. They bought up available chunks of prime waterfront property, sometimes even taking partially submerged land and filling it in. The real-estate market in Port Isabel soared.

What was good business for the developers and condo merchants, and for the condo buyers from Dallas, Houston, and the Midwest, however, was not necessarily good for the local residents. Rents increased, as did prices for houses and lots, and many low-income residents were forced to move, priced out of the town they had called home for generations. Some moved several miles down the road to Laguna Heights, now a poor barrio many of whose residents still work in Port Isabel, the town of their birth.

Port Isabel's limited local resources have also been strained. Fresh water is now a limited resource in the summer months. City services, such as law enforcement, have been hard put to keep up with the recent increases in population. For the first time the boat harbor is showing signs of neglect caused by a lack of local funds. In fact, many residents hotly argue that the quality of their lives is deteriorating even as the developers are planning for and constructing more condos.

On the up side, the condos have created, either directly or indirectly, more jobs. The downtown section of the town is booming, prosperity evident in the new stores that have opened their doors. Someone must feed, entertain, clean for, and care for the condo residents. The increase in service jobs has helped many families of shrimpers make ends meet. The sons and daughters of shrimpers build the condos, cook the food in the kitchens in the new restaurants that cater to the condo crowd, clean their rooms, sell them goods from behind store counters, and provide all the services that a couple from Dallas or Houston would expect and require.

At the same time, the new condo owners pay local taxes, which provide funds for the local school district. Recently a new high school was erected and a new addition to the football stadium completed. The nineteenth-century village is turning rapidly into a twentieth-century town.

Condos are, of course, not cancer, but I still remain more than anxious for Port Isabel. Condomania is quickly replacing a way of life that spanned 130 years, years marked by a dependency on fishing. Fishing is not easy work; it is backbreaking and, at times, dangerous. Year after unending year the people of Port Isabel sweated days and nights for little pay. Change was inevitable and welcome, yet I wonder who really benefited and who did not.

The majority of the people in Port Isabel have had nothing to say about the changes that have come and are coming to their community, changes that affect their daily lives. Even as they are displaced by an impersonal real-estate market, they have no political input and no voice in their own futures. The strain on limited resources, on available land, fresh water, and, yes, on the fishery itself, has yet to arouse the local politicians. They continue to share in the new abundance while the majority of their constituents bear up under the new burdens.

Port Isabel is beginning to resemble other coastal towns. The individual characteristics that made it unique are being replaced by an eminently forgettable uniformity. After all these years, Port Isabel, peopled by some of the most independent Texans in the state, is becoming "quaint."

The issue here is a basic one, one not blurred by a fascination by or longing for the past. Those who have worked in commercial fishing or in related jobs rarely want to return to the drudgery and economic insecurity that characterize them. But there is a difference between the shared prosperity of the shrimping days and the prosperity of condomania. There is a growing awareness in Port Isabel that what is good for the new people in town, those who live in the condos and those who own the businesses that cater to them, is not necessarily good for everyone.

A time will come in Port Isabel's future, perhaps within the next ten years, when the condo economy will take total control, forcing everyone who cannot afford the high prices to move elsewhere. It is a predictable outcome, given present trends, but an outcome far from inevitable. To change the trend requires a grassroots leader, or organization, that embodies the interests of those, the majority, who have been excluded from the decision-making process. As yet, though, no leader or group has appeared to do effective battle with the politicians and the developers.

Not all Texas coastal communities are undergoing the economic changes that Port Isabel is witnessing. While some communities like Port Isabel are booming, relatively speaking, others are going bust. Quietly.

Sea Drift, near Port Lavaca, is one such town. The commercial bay shrimping industry is riding on tough times. Foreign competition has driven prices down, and state legislation threatens tighter control over the fishery resource. Local petrochemical plants have laid off workers. Tourists are attracted to the beaches and docks of other communities that have more to offer. In 1985 times are hard for people in Sea Drift. As jobs continue to dry up, many will move to the nearest urban area that can provide employment.

Not all parts of the Texas Coast are marching to the same economic drummer. Smaller coastal communities, unlike Houston, Corpus Christi, Port Arthur, or Victoria, are undergoing uneven, usually unplanned economic growth and social change. For every community that appears to be thriving, one remains in economic and social backwaters. Even in those towns witnessing rapid growth, like Port Isabel, life is still far from easy.

Perched on my bar stool, I look across the brass and polished wood of the bar. The bartender and waitresses stare intently at the long table behind us at which are seated twenty-five men, boisterous and growing louder after their third round of cocktails before dinner. Several complain about both their drinks and the restaurant. The men are from Houston, part of a sales team meeting here for the weekend. Do I read anger in the workers' eyes? Or perhaps I, myself, have drunk too much and am ascribing too much meaning to the occasion. In either case, the point remains: all Texans, not just a chosen few, should have a voice in determining the future of their own coastal communities.

Martian Fishermen

The docks at Aransas Pass are straight as an arrow. They border the two long channels that flow in from the bay next to the bridge and the ferries that link Aransas Pass to Port Aransas. From the bridge the Conn Brown Harbor, as it is officially named, flashes by in a second. But walking the docks from one end to the other, a Texas mile long, gives a better perspective to the home of the largest collection of shrimp trawlers along the Texas Coast.

Just before noon on a hot August day I walked slowly up to one trawler so that the crew sitting in the shade of the pilothouse could see me clearly. I asked one of the men, first in English, which got me a shake of the head, then in Spanish, if the captain were around. The captain had gone home and would be back later that afternoon. I said that it was very hot out, the shrimper replied that it had been very hot for the last few days, and then I asked how the fishing was.

He gave me a quick once-over, then asked me if I wanted a beer. I said sure and hopped over the railing onto the trawler. He disappeared into the pilothouse and returned with a cold Bud. We crouched on our heels in the shade talking about how many boxes of shrimp they had netted that week, the low price of shrimp that the fish houses were giving, and the generally poor season everyone was having along the coast.

The other two men vanished into the innards of the vessel. Silently, first one returned, then the other. The man I was talking to, the boat's rigger, whose name was Alex, nodded at them and told me they were Juan and Antonio. I said hello, apologized beforehand for my bad Spanish, then continued talking with the rigger.

Alex was from Matamoros. He was twenty-two, single, and had been a student at the *tecnologico* before he ran out of money. He said that he had begun shrimping in Port Brownsville five years earlier and, through hard work and persistence, had become a rigger. The other two, the headers, looked up to him as if he were their older brother. He had gotten them their jobs on this boat.

At first neither Juan nor Antonio would look at me directly. I, too, avoided making eye contact or appearing to stare at them. By the second round of beers they had inched closer in the shade to hear the conversation. I told the rigger, as much for their benefit as his, that I was a researcher at the university doing a study of shrimpers. He seemed impressed; so did the headers. I told them I was studying shrimpers because their work as fishermen was sometimes both difficult and dangerous. The rigger smiled and nodded. Juan and Antonio visibly relaxed.

The heat grew oppressive. We went inside and sat around the galley table. They brought out more cold beers. Juan told me this was his first trip out as a header. We talked for a while about being seasick, and Juan and I laughed. He said he had been sick for just one day, and I told him about when I had been seasick for almost a week straight. We agreed that there was nothing that compared to being seasick.

Juan was seventeen. He had worked at odd jobs since he was thirteen. I asked him why he wanted to give shrimping a try, and he answered readily, "The money." A header could earn, he had heard, up to five thousand dollars a year if he worked hard and had some luck. I asked him what he was going to do with all that money. He said he would save it, then decide what to do with it after he had time to think about it.

Juan wore jeans cut off at the knees, ragged at the edges, a blue T-shirt with the sleeves cut out, and blue thongs. His hair was cropped short, and his eyes were still bloodshot from the nights spent working in the Gulf. He had a crude tattoo of a dagger on the inside of his forearm and a small gold cross and chain around his neck. Juan was friendly, outgoing, and enthusiastic about his future. He intended to stick to his job for at least a year and, if he liked it, maybe become a rigger like Alex.

I asked him what he thought of Aransas Pass. He said he knew nothing of it. Like Antonio, he had stayed on the trawler since it had docked, and he would not leave it before the next three-week trip out, maybe tomorrow or the next day. I asked him if he was bored. He said he was, but soon he would take a break from the boat to visit his family. For now, all he had to do was rest up from the last trip and help Alex clean up the boat.

Antonio sat on the hard wooden bench under the porthole listening intently, drinking his beer and smoking an occasional cigarette. Antonio was more like Alex than like Juan. Educated, well on his way toward becoming a well-paid electrician, he had quit school when his family needed money. Like Juan, he lived at home. Like Alex, he wanted to get married after he had saved some money and met the right woman.

Antonio viewed fishing as strictly a meal ticket. He did not mind the hard work, but the life of a shrimper was not for him. He would work a few more months at it, then he hoped to find a job in Matamoros. He said that a job at one of the American factories was opening up, and he had some friends who would help him get it. In the meantime he was making the best of an unsatisfactory situation.

Alex interrupted regularly to inform me that Juan and Antonio were new to the boat and did not know much about fishing. He, Alex, on the other hand, had shrimped for five years and, before that, had worked as a fisherman in a small community not far from Matamoros. He had fished regularly with his uncle and one of his cousins in the Mexican bays south of Port Isabel. Occasionally, when the Gulf was calm, they would venture out into the sea to catch larger fish.

Alex said that he was a real fisherman and that Juan and Antonio were just working at it for a short time. Juan and Antonio agreed. I asked Alex if the trawler captain was also a good fisherman, and Alex said he was average. Sometimes the captain would bring in three to five boxes of shrimp a night, but for the last six weeks they had been netting only one to two boxes. Alex did not seem upset, just rather matter-of-fact about their bad luck.

Juan offered me some food. I said no, thanks, as graciously as I could. Juan got up, went to the refrigerator, then turned to the small gas stove where he worked up tortillas, eggs, and *chorizo* for himself and the other two. More beer appeared. I declined the extra beer, saying I had to go talk to some more shrimpers and that the beer was already making me sleepy. We talked on through the meal. We heard footsteps outside on the dock, and Juan quickly jumped up and went out the door to see what was going on.

A few seconds later the captain walked in. He was surprised to find me on his boat. I explained who I was and told him we had been sitting around having a few beers and talking about shrimping. I asked him if he had some time to answer a few questions. He said he did. We went up forward behind the wheel. From the moment he had entered, Juan and Antonio had stopped talking, and Alex, who had been speaking in beautiful Spanish, switched to his broken English.

The captain, named Gus, was a native of Aransas Pass. He had an uncle who had fished out of there since the early 1950s. Gus, despite being only in his middle twenties, had been a shrimper for almost ten years. I asked him the usual round of questions, then turned to the subject of his crew.

Gus told me, matter-of-factly, that Alex was the best rigger he had ever had. Alex was dependable, easy to work with, and knowledgeable. Gus said that Alex's only drawback was that he spent too much time teaching a new header the ropes—only to have the man leave for another boat or go back to Mexico.

Gus spoke a little Spanish and said he and Alex had no problems communicating. Gus liked Mexican food, having been brought up in Aransas, so Alex's cooking suited him fine. He said the only thing that was a hassle with a Mexican crew was their having to sneak around all the time like they were criminals.

"They work real hard when we're out in the middle of nowhere, then when we come in to tie up they have to hide out. It's stupid."

I asked him if he would hesitate before hiring another Mexican.

"I don't care where they are from. Just so they can do the work," he said quickly.

I thanked him for his time and returned to the galley, where I found Alex and Juan still sitting around the table. They had hooked up a little electric fan and were trying to stay cool in the heat. I thanked them for the conversation and the beer and wished them good luck in their fishing. When I asked where Antonio had gone, they said he was sleeping.

Climbing over the rail and onto the dock, I walked through the stultifying heat to my car, opened the doors to air it out, then got in and turned on the air conditioner full blast. I drove over to the main business district, found a Whataburger, and squeezed into a booth with a large iced tea.

I spent six months interviewing shrimpers from Brownsville to Freeport, and it became clear that a considerable percentage of the work force was undocumented workers from Mexico. A smaller percentage came from farther south. I had already interviewed two men from Belize, one from Nicaragua, and several from Honduras.

About 20 percent of all the Texas shrimpers I interviewed were "illegal aliens." I don't like that term, although it is always used. It makes the men sound like criminals from Mars. It seems a demeaning term, especially when applied to men like Alex, Juan, and Antonio. These men were good workers, asking very little from their employers for what they gave in return. They were friendly, homesick, and tired of hiding out.

I had expected to find undocumented workers from Brownsville to Corpus Christi, but it soon became evident that they were working out of Galveston, Port Arthur, and other northern Texas ports as well. In fact, captains often preferred Mexican crews because they had a good reputation for doing the work, complained little, and were easy to get along with. The irony, of course, was that in spite of all this they were, in the eyes of our judicial system, criminals.

It is often said that undocumented workers take jobs away from American citizens, in this case Texans. While there is no real labor shortage in the fishing industry, there is a lack of experienced crewmen who not only know their job but also have the personality that makes them tolerable to others during the long hours at sea.

Mexican workers were ready and willing to work on the shrimp trawlers. Most did not consider shrimping a lifelong occupation, but the money they could earn in Aransas Pass, for example, was far better than that in Matamoros, Tampico, or Veracruz. The big trade-off, however, was working in a foreign culture many miles from their families and their homes. The ones who stuck to it could earn between three thousand and six thousand dollars for six months' work, more than three times what they could expect to earn, if they could find a job, in Mexico.

So the men are not taking jobs away from Texans. In truth, it seems more appropriate to discuss how they are helping the Texas shrimping industry. Simply put, without them, profits to trawler owners would be lower and supermarket prices for shrimp would be higher. When Texas shrimpers are feeling the economic crunch from all sides, the Mexican contribution to the labor force should be seen as a blessing.

When I first came to the border a decade ago, I held a number of common misconceptions about undocumented workers. Now researchers have laid to rest a majority of those ill-founded charges. Among maids and mechanics and shrimpers, my own experiences support what the researchers have found.

But whatever the facts, the issue is far from simple. Thousands upon thousands of workers continue to enter this country illegally, crossing the Rio Grande in search of jobs. Undoubtedly illegals in some occupations—farm workers, for instance— contribute to low wages and threaten unions and unionization. The vast majority of undocumented workers are, however, like Alex, Juan, and Antonio; they do their work well, then return to their homes south of the river. The difficulties remain, and, as I write, yet another version of the Simpson-Mazzoli Bill is defeated.

I finished my second iced tea, climbed into my hot car, and drove back to the port. I circled the first channel and drove to the end of the road. There stands a three-story white stucco tower with a huge cross and a Christ who faces the small harbor. There is a low chain-link fence around the structure. Two large buoys lie in the grass on either side, and two small palm trees flank the buoys.

Inside the building a black wrought-iron chandelier hangs from the ceiling, and a wooden crate looks as if it protects a piano or an organ, or perhaps a simple altar. Otherwise the single small room, open to the elements, is bare.

Outside on one of the tower's faces are two granite tablets, one large, one half its size. The smaller one is dedicated to six Coast Guardsmen who died trying to rescue a shrimp boat. The larger of the two reads, "In memory of fishermen who lost their lives at sea."

The tablet is filled with more than sixty names. The names begin in the upper left-hand corner and stretch, in two columns, to the lower right corner of the tablet. There's no more room for names on it. Circling the tower I found a new granite slate, of the same size, erected since I had first visited the memorial two years earlier. This newer tablet is completely blank, awaiting its first names.

I wondered, looking at the empty slate, how many Mexican fishermen died and would die while working on Texas trawlers. And when they did, they would still be called illegal aliens.

The next morning I drove out to the port. I parked near the Gulf King complex, one of the largest fleets in Texas. From a distance, the buildings surrounding the docks looked like large Iowa silos. I walked over to the docks, searching for the trawler I had been on only the day before. But in its space another boat now lay tied to the pilings. I asked one of the crew on board if he had seen the shrimpers I was looking for. He said he had. He had talked briefly to them as Gus pulled his trawler slowly away from the dock and headed seaward with Alex, Juan, and Antonio. They would be back in three weeks.

The Ones That Get Away

Nature gives out warnings, sometimes small warnings. Nature first began warning me one late fall afternoon about the limitations of my ability to catch, lure, net, hustle, or otherwise conjure fish into my freezer. I sat fishing from a boat dock on the "fingers" section of Port Isabel. I had tried everything in my bait box and, somehow, finally lucked onto a nice flounder, just about the size necessary for breakfast for two.

The luck made me greedy. I sat praying for a second fish as I bounced the lure off the bottom of the tiny boat channel. My fishing partner Mike had been struck by the same greed. We both ignored the fact that for the last three hours all we had caught were ten salt-water catfish, which we had released in disgust and frustration. I have never found a way to satisfactorily cook up saltwater cat, although I have heard many stories of how good it is supposed to be if prepared properly.

Adrenaline surging, we concentrated on some serious fishing. That is when Nature decided to give me her first small warning. She conjured a sea gull to fly over me at an estimated twenty-five feet and to defecate. The gooey mess landed on the right side of my head and completely covered the right lens of my prescription sunglasses.

With my good eye I looked helplessly over at Mike. He sat not three feet from me, hands tensed around his rod, eyes on the water, waiting for Moby Dick to strike his line. He remained untouched. With a certain degree of stealth I cleaned up what I could, fearful he would see me. That was when I felt the tug. Not just any tug, but the tug of a substantial fish. It was no catfish, and certainly not a flounder, at least not any flounder I had ever netted before. I thought small shark or mystery fish. I settled on mystery fish as I cautiously began reeling it in.

Nothing hurts a fisherman as much as maneuvering the fish close in, getting a glimpse of it, then seeing it disappear forever. So I was very careful. No loud yells to Mike for help. Just determined arm muscles, the brain setting up scenarios of all that could go wrong in the last fifteen seconds before the fish is landed. The sea gull was temporarily forgotten.

I bullied the mystery fish close in. Still no glimpse. With a flick of the wrist I jerked the tip of the rod and the mystery fish flew out of the water and onto the dock with a strange scraping sound. A pair of prescription glasses, slightly barnacled, entangled in the lines, weights, sinkers, and hooks of maybe three other fishermen, lay on the boards before me. The contraption weighed at least four pounds.

I walked over to the bait shop at the end of the rickety pier and ordered up a beer. I returned, cleaned the remainder of the gull mess off my glasses, drank the beer, and stared at the glasses fish. Clearly Nature was giving me a message, first by gull, then by imitation fish. She was telling me it was time to hang up my rod and reel forever.

I really did give fishing my best effort. I fished the bays, the surf, from docks, from boats, the whole bit. And I did catch fish, although never as many as I would have liked. I began my serious Texas fishing off a pier in the Gulf. At night. I used live shrimp and caught sheepshead, big ones. Sheepshead are like dumb ox. I stood hovering over their fish shadows under the pier light and gingerly stuck live shrimp into their mouths. Most of the time they just stared at the bait, even when bumped on the nose with a delectable delicacy doing the old soft shoe. Sometimes they would take the shrimp into their mouths, then spit it out. The skills involved are not exactly those required when trout fishing in a Colorado mountain stream, but it was all I knew. Anyway, the real problem with sheepshead is that you need a chain saw to clean them.

Night fishing can get strange, especially when there are not many people around. Off the T-heads in Corpus, or the old piers around Lavaca, or the ones that jut out into the Laguna Madre from Port Mansfield, there is almost always a stiff breeze. The night wind and surf muffle any human sounds. The ocean becomes a backdrop to a fisherman's concentration. Rather than just stare at the dark waters, a fisherman will start thinking about what the fish are doing and why the hell they are not biting. Or start fantasizing and, after a few hours, hallucinating. Any free-floating daydreams, nightmares, or other neuroses that happen to be hanging around may surface, along with an occasional fish.

When and if the fish do start biting, it is frenzy time. Depression skyrockets into elation as the fish keep taking the bait. Driving home at four in the morning with an ice chest full of sand trout is definitely addictive.

My ideas about fishing were already beginning to change before I was dive-bombed by the sea gull. One golden morning in May I was fishing the bay with my friend Genaro. Arriving just after the sun, we had waded out into the water a good quarter mile. The expanse of bay was crystal clear, like a large New England glacial lake. On the leeward side of Padre Island there was not a breath of wind. I could see my Japanese sandals clearly as I trudged through the knee-deep water, alternately weeds or sandy bottom.

We were using bright silver lures. When we cast, they sparkled briefly in the new sun just before they touched the water. I saw the silhouette of the trout as it circled Genaro's lure but passed it by at the last second. He cast again, in the same direction; again the lure with its treble hooks gleamed momentarily before it hit the bay waters. This time the trout rose up from the weed bottom and struck the spoon. Genaro waited the brief seconds to be sure the trout had taken the hook, then with a flick of his wrist he secured the hook in the trout's mouth. The struggle was brief but noisy in the quiet bay.

By the time I waded over to his side, Genaro already had his catch on a yellow line tied to his belt. He pulled the fish out of the water, and I admired it. It was a good two-and-one-half-pound sand trout. I touched its underbelly with my fingers. It felt flossy, like a photographic print, still wet, removed from its developing tank.

We fished for several more hours. I caught two fish but spent most of the time staring into the water at my feet. I could see the shadowy outlines of rays before they skittered away. I had shuffled my feet a few yards from them and watched them glide their barb-tailed bodies into safe hiding places in the weeds. The flounder were an even odder sight. If I concentrated really hard I could pick out their shapes just below the sand, where they hide waiting for food to drift by. I tried to imagine where their two eyes sat, side by side, watching patiently for lunch to come along.

The trout swam around us in small schools of four to five fish.

They covered an incredible amount of space in their wanderings. One school would pass by, only to be met by another headed in an opposite direction. Then the original group would circle back. Occasionally a grey shape sped by, unidentifiable.

Around ten o'clock the wind began picking up. The waters turned slightly choppy, once again opaque to the human eye. We retraced our steps, iced the fish, and drove home. Riding home, I realized that I had enjoyed watching the fish more than catching them.

Working on the shrimp boat finalized the process. I spent one afternoon fishing off the stern of the trawler, fifty miles out in the Gulf. With a rope, a half-inch hook, and a beer bottle for a reel, I fished while the rest of the crew slept. I fell into a school of dorados. For a half hour I knocked them on their noses with every piece of bait I could think of, from two kinds of shrimp to squid, pieces of small fish, and back again to shrimp. Nothing, not a single bite.

Every five minutes or so I would think about giving up. The dorados were like sheepshead, only more tempting because they were bigger and there seemed to be hundreds of them. I waited. A gull decided to dine on the fish bait I was using. I chased him away, frustration mounting. Here I was in virgin fishing water surrounded by more fish than I had ever seen in my life, some of which looked from my vantage point to go at least eight or ten pounds, and all they were doing was their best impression of a sheepshead. I waited.

Suddenly, for no reason that I could discern, they began to bite. I set the hook, then jerked the fish up out of the water and over the gunwale, the way I'd seen tuna fishermen do in an old television movie. The fish lay flapping and hopping on the deck, at least a four-pounder. I baited the hook again, and it barely touched the water before another dorado sunk its teeth into it. Again, I flipped the fish over the side of the boat.

This went on for about twenty minutes. Soon I had a large stack of dorados on board. I began being selective about which ones I fished for. I carefully ignored the smaller shadows below me, lowering my bait directly over the largest ones I could see. Once the hook hit the water, it was immediately gobbled up by the nearest fish. The biggest dorados were soon on the deck behind me.

My arms began to give out with the effort, but not before my desire to fish had completely disappeared. I gave the fishing gear to the header, who had heard the rumpus and, half asleep, wandered over to see what I was doing. He spent the next half hour finishing off the school of dorados. Between us we had caught more than forty fish. I should have been elated, but I felt just the opposite. Catching the fish, after about the fifth one, had become meaningless. It was too easy. There was no sport in it, no skill.

I looked over at the still-shiny bodies of the fish, any one of which, on another fishing trip, I would have considered a trophy. All I could remember, however, was their silky motions as they schooled off the trawler, rising and falling into the depths in their constant searching.

I discovered on the back of the shrimp trawler, as I had on that morning with Genaro, that I far more enjoy watching fish than I do catching them. Alive and moving they are incredible creatures; in their motion is their beauty. When they are dead or dying, all I am left with is the story of their capture or their uniqueness before they took the hook. I certainly do not begrudge others the pleasure they derive from fishing; I have simply taken nature's small warnings to heart. My rods and reels are busy gathering dust. The last fish I ever caught was a pair of prescription glasses.

On Not Sailing

The sailboat sits under a friend's stilt house. Against one of the twelve thick posts that support the small structure, the trailer rests, its balloon tires wedged into the sand. Rains during the last months have washed over and around the rubber up to the wheel bearings. The dune grasses from the lot behind the house have crept up under the trailer. The mainsail is turning a pale green in places, even though Dacron is supposed to be immune to mildew. In the cockpit the wooden benches are encrusted with layers of salt and grime. It is my sailboat. I have not sailed her for nine months. No, that's a lie. It has been almost a year. It's killing me.

This was not supposed to happen. For years I made fun of others who would buy a boat, sail it a few times, then let it decompose in their backyards. I worked out a simple formula based on observation: the bigger the boat, the less it was used. There was definitely an inverse relationship between the money put into a boat and the time it was actually sailed. Of course, this was common sense in a way. A person with a bunch of money seldom has a bunch of time. So he buys a symbol of his wealth, a sailboat, then never has the time, after the initial enthusiasm, to sail all thirty-two or forty feet of it. The boat begins to show signs of serious neglect. He avoids it, riddled with increasing guilt. Finally, shamefaced, he sells it, maybe three or four years down the line.

By then he's taken a loss, but the guilt has been removed, and he feels true relief as the new owner proudly sails away. He no longer has to bear the sight of his rusting, rotting treasure tied to the dock. If he's smart, he waits at least a year or two before buying another big toy.

Having seen the phenomenon repeated from Houston to Brownsville, I sought to escape it. I bought a small sailboat, under eighteen feet, that could be left in the water or trailered. The sailboat was simple in design but engineered to withstand the bumblings of an inexperienced sailor. A Boston Whaler 5.2, she draws less than six inches with the centerboard up, making her excellent in Texas bays. She has plenty of mainsail, a good jib, wood in the right places, and a broad beam that holds all the family and then some. I read the literature until the pages of the brochure began to crumble. The boat was exactly what I wanted, large enough so I would not be tempted for some time to "trade up," and constructed to last, so that the running gear would not expire at the same time as the bank note. I talked it over with my wife many times. At the last second I got cold feet; after all, I had never even owned a boat before, let alone a sailboat. But I bought the boat.

By the end of April, even though the bay winds were strong, gusting up to twenty-five knots, and the waters were still nippy, I already had sailed her five or six times. That first summer I sailed two or three times a week. I was not teaching summer school, was between research projects, and had decided that sailing deserved my full attention. Some friends helped me rig her.

I kept the boat at home in my driveway and hauled her out to the bay with an old 1948 Dodge truck, a three-quarter ton with a huge trailer hitch on the rear bumper. I had the truck's engine rebuilt and, although it looked as though it would fall apart at any moment, it ran smoothly and quietly. The Dodge was the kind that had three rear windows in the cab, one large one in the center, and two on either side that wrapped around the corners. I would stare through those wraparound windows at the trailer and boat attached to the rear of the truck every chance I got as I drove along the road. For many months I could not believe what I saw way back there, extending twenty feet beyond the truck's bumper.

I now had the unbelievable choices, which I relished, of putting my boat in at the beach, or off a ramp, or just unloading her gingerly into the bay. If I wanted to just sit on the bow and look at the sunset, I could. Or I could paddle around, or sit for hours in the middle of the bay. These were my choices, and the choices were exhilarating.

Initially, the preparation for leaving my house took about an hour and a half. First I had to load all the gear into the truck, including the tiller, which was the most awkward piece of equipment. Having mentally noted that the jib, life jackets, tool chest, and extra running gear were all in their places, I started on what I called the secondary equipment, the ice chest, ice, Gatorade, sunglasses, and all the little stuff that, at the right time, made a huge difference. Once everything was in place, I hooked up the trailer to my truck and tugged the trailer effortlessly out of the driveway, onto the street, and toward the bay. That old truck could have trailered an additional three thousand pounds with no trouble. The truck had an extra low gear that had incredible power.

The main drawback to the truck, ignoring the fact that it had no air conditioning, smelled like a barn, and still had the original bench seat, was that it took an hour and a half to cover the twenty-five miles to the water. In the beginning this was no problem. I had the whole summer and was in no hurry. I did not want to push the truck past forty. I had time to think pleasant thoughts during the trip. I drove on the shoulder of the highway, ignoring the traffic as it sped by. Not a few who whizzed past admired either the truck, the sailboat, or both. I was a proud, if inexperienced, sailor.

Usually I put the boat into the water on a small stretch of sand between a marina and a restaurant. After a few tries I learned how to back the trailer in. Then it was just a matter of putting the trailer into the water up to its bearings, unloading the boat, and doing the final rigging. The best moments, however, were rigging the boat, including stepping the mast, while she was still on the trailer. Parked off to the side of the small beach, I lazily set about raising the mast, attaching the boom with a single long bolt, and attending to the other little details that require concentration but no great expertise.

There was almost never a time during this rigging, however, that a problem did not arise. Often the problem was of my own making; more rarely it grew out of the nature of the boat itself, or perhaps of the trailer. I would forget an important piece of hardware, or a line would jam, or I would discover rust on a cleat that threatened to do some minor damage. I made the adjustments, polished the brightwork, solved the small problems. The pleasure of doing so sometimes made the sailing itself less important. In devising a new way to raise the mast without a certain pulley, or discovering a cracked fitting before it broke, I not only came to know the boat and its peculiarities, but I also thoroughly enjoyed the time I had doing it. I would punctuate my work with frequent trips to the Dairy Queen down the street for iced tea. I walked slowly to and from, absorbing the heat of the day, feeling the breeze on my bare back and chest.

I became a better sailor. Not great, but good enough to feel confident. When I finally had winched the boat onto the trailer, trucked home in the fading light, hosed truck, trailer, and boat free of salt, and stored away the gear, I collapsed on the sofa, filled with the experiences of the day. My arms ached, my feet usually had new cuts from the hidden oyster beds, and often I had a double sunburn, but the sail had cleansed me, washed away the impurities of living on crowded land. The next morning, stiff from the day before, I completed the few small chores that remained, like emptying the ice chest that I would forget or skip in the dark in my front yard. I almost always found a piece of gear stuck in an unlikely place, often in the anchor locker in the bow of the boat. That summer, full of sailing days, sped past.

I sailed, though less frequently, through September and early November, the blue northers finally putting an abrupt halt to my first season of sailing. I hauled the boat into the backyard, covered her with a canvas tarp, and waited for spring. In late February, the sailing urge running deep, I tore down all the mechanical hardware, cleaned and oiled it, attacked the rust that lingered throughout the gear, and reassembled the rigging from scratch. I now felt that I knew the boat from bow to stern. I liked her.

Then followed three years of sailing. Each year I went out less often, not because I had lost interest or enjoyment, quite the opposite, but simply because my life became more complicated. My time became less my own; children, research projects, and business interests all made it more difficult to take afternoons off. The Dodge truck died, overtaken at last by old age. Unwilling to face its death, I let it sit on the street in front of my house. Its bed filled with debris from the neighborhood ebony trees. Plants appeared on its running boards and tailgate; tiny ebony saplings grew along with sprigs of Johnson grass. Someone suggested that the Dodge was beginning to resemble a large gold flower pot. Finally an acquaintance asked if the truck was for sale. I said yes, and in two weeks he hauled it away, filled with the pride of new ownership.. I saw the truck six months later. He had painted it a cherry red. The old Dodge had been given yet another reprieve from the junkyard.

Without a truck, hauling the boat to the bay became a problem. I had a hitch welded to the small frame of my wife's Honda Accord, but the results were less than satisfactory. The Honda strained under the load; three times in a row I got the car and trailer firmly stuck in the beach sand trying to pull the boat out of the water. Each time a friendly passerby stopped and, with his four-by-four or van, muscled the boat and trailer out of the ruts I had made in the sand. Still, getting stuck would have been much less trouble if I had had the time to think about the best strategies for removing the boat from the water. But there were articles to write, babies to be watched, bills to be paid. I grew impatient.

I stopped hauling the boat from my house to the bay. In the Honda I was constantly worried that I would either tear the transmission to pieces or lose control of the trailer and boat, jackknife, and have to paste the pieces back together. It never happened, but the anxiety and worry made me enjoy sailing less. Once on the way to the bay I did lose a tire from the trailer. It went bounding away into the salt flats, and the trailer axle dug a long trail three inches deep in the asphalt before I could stop.

I arranged to store the boat underneath my friend's house, only three blocks from the bay. That way I could jump in my car, motor out trailer-free, hitch up the trailer, then rig the boat and put her in the water. But the commitments increased. Some weeks in the summer I had no time at all to get to the beach, not even for a walk. When I did, I found often that I was too tired to enjoy the idea of a quiet sail. The boat sat. The more she sat, the more the elements got to her. She began to show the telltale signs of neglect. I began to feel guilty when I went to the beach and did not sail. The guilt made me feel resentful. It was my sailboat, and if I chose not to sail her, then that was my choice. The resentment led to more guilt. The guilt increased as the salt air began to corrode the hardware, mildew the lines, and pit the wood.

Unfortunately my favorite restaurant on the island sits directly across a sand dune from my rusting boat. Call it a quirk of fate. I sit on the sundeck of that restaurant, Blackbeard's, on occasional afternoons with my back to the boat. It is all very confusing. More involvement in life, which brings more enjoyment, leaves less time to enjoy. Sailing requires time, enough time not to be concerned with the problem-solving that is a part of the sailing experience. Sailboats by nature are not the kind of vehicles that one can dash out to, jump into, and sail away. Their identity requires appreciation; their form demands an intimate knowledge.

I have, at least for the moment, assuaged the guilt by making myself a promise. I will not rush again to sail. I will buy a truck or van to do the hauling. I will not sell my boat. I will let the boat rust until I have the time to set her right.

The Hearing: Black Balloons

If it had not been in my town, my home, I know I would have ignored the public hearing. I normally avoid hearings like the plague. In part I do so because of my distrust of how large bureaucracies tend to function, a distrust based on both my own personal experiences and a more theoretical knowledge gleaned as a sociologist. In part, I do so, too, because hearings always verge on the extremely boring, even when the subject is crucial.

This hearing was different. It was about the burning of carcinogenic wastes, three hundred thousand tons of the stuff, in the Gulf of Mexico about 170 miles east of Brownsville, on the border with Mexico. The sheer mass of the amount to be destroyed boggled my mind, and the site, given the amount, seemed much too close to home. What if the dangerous junk wasn't completely destroyed and floated into shore? Or what if the ship sank?

A company called Chemical Waste Management (CWM) wanted permission from the Environmental Protection Agency (EPA) to destroy the hazardous chemicals. CWM's ship, the *Vulcanus*, would begin regular burnings if the permit were granted. Already the EPA had granted permits to the company for two test burnings.

As if this were not enough to shake up my usual apathy about public hearings, the event was to take place less than a block from the college where I teach. Several of my students were going to testify, and a student-supported protest was planned. I waited with anticipation for the day of the hearing.

November 21, 1983. A Thursday, warm, windy, and the sun peeking steadily from behind clouds all day. The EPA officials, Kuntz, Sussman, Oberacker, Jackson, Schatzow, Alan Rubin, and Asistant to the Director Jack Ravan, all sit behind long tables at the front of the stage. They look out on an almost-empty Jacob Brown Auditorium, even though more than five hundred people are present.

At nine the public hearing is opened by Mr. Rubin. He and others on the panel speak for an hour or more on the arrangement already made between the EPA and CWM to safely burn millions of tons of PCBs, dioxins, and other toxic wastes in the waters off Brownsville. Their explanations and renditions are met by boos, laughter, and sarcastic questions that they ignore. The boos echo in the cavernous hall.

About ten o'clock the local politicians approach the lectern on the stage. To the delight of the audience they seriously question the granting of permits to CWM. But the articulate oratory of one state representative, the "good old boy" talk of the other, and the surprisingly emotional and moving speech of an official from South Padre Island elicit no response from the men on the stage. Each politician receives a curt "thank you" and is dismissed promptly and efficiently. I wish that at least once the men from the EPA would grimace, laugh, show some kind of human emotion. Charged with incompetency, ignorance, and insensitivity to local needs, they show blank faces.

After long applause, some of those in the audience leave their seats to shake the hands and thump the backs of their elected representatives. The politicians stay to mingle, be seen, and show their support for the diverse audience.

I recognize doctors, a surprising number of lawyers, business people, school children (some carrying lunch in a brown paper bag) from the public and private schools, and teachers and administrators from the college. The environmental groups are here, along with their detachments of legal experts and overdressed housewives. The audience is predominantly white, upper-middle-class. This is an area where 80 percent of the population is Mexican American and poverty the rule, not the exception. The audience grows increasingly angry at the men on stage.

Squeezed between audience and stage, the press stands and sits in and around two tables. The cameramen, young men in their early twenties, toy with their gear when not filming or talking with each other. The television reporters, mikes in hand, sit wearily in their chairs, puzzled by their central problem: How does one film a spontaneous event? More at home in the controlled setting of the press conference, the interview, or the event purposely staged on their behalf, they grow increasingly troubled by their inability to cover the hearing. From their perspective, there is far too much boring talk, no action.

An articulate speaker who does thrill the audience often finishes before the crews can turn on their lights and cameras.

On the other hand, the newspaper press, both the locals and those from Houston, Dallas, and Austin, filter through the assembly. Among the wealth of politicians, expert witnesses, environmentalists, and school children they find their stories.

The chanting begins at about 11:20 in the morning. Two hundred college students file into the back galleries of the auditorium shouting, "The *Vulcanus* burns me up." The cameras roll. Each student holds a balloon in his hand. The balloons float on strings above the students' heads. The balloons are a dead, flat black. One or two escape and rise slowly toward the rafters, some forty feet above the heads of those on stage.

The chanting continues as two of their leaders, one an English teacher at the college, the other a student who two days earlier had told me he would miss my sociology class because of the protest, came forward.

The English teacher approaches the lectern first and, in very polite terms, denounces the burning. The students applaud. The student leader then steps forward and speaks, briefly and haltingly. "I represent all the students at Texas Southmost College when I say we are against the burnings. That's all I want to say." The students applaud wildly. To the delight of the gallery, the two leaders then spread before the audience and the cameras a large banner that reads, "The Vulcanus Burns Me Up." The applause lasts a half minute, then the two walk off the stage.

The students in the galleries begin leaving almost immediately, their demonstration at an end. For almost all of them it is their first political protest. Mexican Americans, they are poor; often their involvement beyond the college is limited by the part- or full-time work they need to support their families and themselves. As one of their teachers, over the last ten years I have come to realize that their respect for and fear of any representative of authority made it difficult, if not impossible, for most of them to become politically active. That so many had voluntarily participated in the short march and brief demonstration surprises me. I imagine they did so in spite of opposition from their friends and, perhaps, against the wishes of their families—if their families are even aware of their actions. (The next day, when I have a chance to talk with many of them, I find my suppositions to be accurate. Considerable planning and effort went into their march; the students hope that it was effective and seem relieved that it is all over.) I think back, quickly, to my own participation in the marches of the sixties. My own fears, like my students', arose from a feeling that arbitrary decisions were being made that had long-term implications for my life. The feelings are the same; only the circumstances have changed.

What I know, then, was a considerable effort on the part of the marchers is quickly dismissed by the men on the stage. They see the protest as brief, overcourteous, and ineffectual. The initial chanting gave them cause for alarm, but a few hundred well-behaved college students are not something to worry about. When the chanting subsides, the hearing officer quickly calls the next speaker to the microphone. The black balloons leave the auditorium, and the business of the hearing proceeds as before. High above in the rafters of the auditorium two black balloons are all that remain of a failed attempt to change the collective mind of the Environmental Protection Agency.

The Hearing: Leper Ship

I wish I had been there that evening to see the faces of the men from the Environmental Protection Agency when Valley Interfaith marched into the auditorium, fifteen hundred strong and more waiting outside. Valley Interfaith wanted the EPA to prevent CWM from burning toxic wastes in the Gulf of Mexico. From all over the valley they had come, representatives from each parish church carrying signs identifying themselves. They marched in a candlelight parade from the Catholic church in downtown Brownsville to the Jacob Brown Auditorium.

The rest of the day at the hearing had passed uneventfully, except for the appearance of Mark White, governor of Texas. Right before lunch there had been a commotion at the rear of the hall. From where I stood, fifteen feet from the stage, I saw a small group of suited men enter in a flurry, shake hands like crazy and, in return, receive growing applause from those around them.

Mark White climbed the steps to the stage and, for twenty minutes, took charge of the hearing. He observed that the people of Texas had little trust in the EPA, given the recent scandal involving Rita Lavelle, former director of the agency. He noted that she had been charged with diverting superfund monies from their intended purpose (cleaning up toxic sites around the nation) to the campaign coffers of his recent opponent. The whole mess was called "Sewergate," and White did not think much of it.

So here before them stood living proof of Rita's lack of success, and Mark White, governor of Texas, made it clear he was going to give these city slickers from Washington a piece of his mind. The audience roared. The speech continued, the governor comparing the *Vulcanus* to the *Titanic*. When he finally ended, Assistant to the Director Ravan jumped up from his chair to shake the governor's hand. As the governor was making his exit, Ravan came to the microphone and assured him and the audience that his agency would closely and fairly examine the issues that had been raised. With that he called a recess for lunch. The crowd, buoyed by the governor's speech, slowly broke up.

For twenty minutes I had actually felt that the hearing was really going to change the opinions of the EPA, that they would restudy the issues and cancel the permit to CWM. So, I think, did many in the audience. But after lunch it was back to business as usual. Witnesses came and went, from housewives to environmentalists to scientists, with no apparent reaction from the EPA officials. The hearing was adjourned for dinner. That is when the men from Washington returned to find the auditorium packed with angry, well-organized protestors.

Valley Interfaith, sanctioned by the local Catholic bishop and created along the lines of community self-help organizations in San Antonio and Houston, was fresh from several recent political victories. Composed of middle- and low-income parish members, predominantly Mexican American, led by local leaders trained by professional community organizers, Valley Interfaith was the first political organization to threaten seriously the existing power structure in the valley.

Valley Interfaith had won funds from Governor White to create a health-care facility for the indigent. They planned to create a regional vocational-training school for high school students. They pressured federal agencies for relief funds following the "Big Freeze," which heavily damaged the area citrus crop and put thousands of farm workers in the unemployment lines. In their brief existence they had met considerable resistance, but never defeat.

One by one, men and women stood at the lectern and, in emotional statements, pleaded, cajoled, reasoned, and sometimes threatened the men on stage. Their leaders demanded to know, finally, what the EPA decision would be. Assistant to the Director Ravan replied sarcastically that it was against the law for him to answer that question. Finally ruffled, undoubtedly tired of the criticism, the verbal abuse, the constant questioning of his agency's credibility, Ravan revealed a semblance of human emotion, anger, behind the mask of the rational bureaucrat.

That night, when it finally turned ten o'clock, the leaders of Valley Interfaith vowed to continue their struggle against Chemical Waste Management, the *Vulcanus*, and the Environmental Protection

Agency. The bureaucrats from Washington, exhausted, returned to their hotel rooms, a few hours' sleep, and then an early morning flight to Mobile, Alabama, proposed site for the loading of the Vulcanus.

I came away from the hearing pessimistic about the future of the burnings in the Gulf of Mexico. Since the hearing my pessimism has not diminished. Valley Interfaith certainly had made their opinion known to the EPA, and the parade of scientists, politicians, including the governor, and local residents had at times been impressive, but I seriously doubted that the various efforts had made any real difference. The general impression I received from the men on stage was that they had already made up their minds and were simply going through the motions prescribed by law.

I was also puzzled by the hearing, puzzled by some problems that I had neither heard discussed nor seen in print. The first was mention of a previous relationship between CWM and the EPA. A little digging in the library revealed that several journalists attempted to show a conflict of interest on the part of the EPA in its treatment of CWM, but no formal charges were ever made.

The reputation and competency of CWM, however, are no longer in doubt. At this writing the corporation that desires to burn millions of tons of hazardous wastes in the Gulf of Mexico has been fined one million dollars by the EPA for improper storage of and accounting for waste materials at its Alabama facility, the same facility that is to serve as the loading site. In addition, on January 24, 1985, the EPA announced a seven-million-dollar judgment against CWM for violations in Ohio involving inadequate storage of PCBs at fourteen different land sites as well as improper sale of carcinogenic wastes. CWM plans to appeal. The company is also, at this time, charged with multiple indictments in nine other states.

Yet there is an even more basic issue than the ability of this particular corporation to honestly and safely dispose of dangerous wastes: Why burn carcinogenic wastes at sea in the first place?

The Vulcanus is called the "leper ship" by those who regularly sail the waters of the Gulf of Mexico. Tankers give her wide passage, their pilots fully aware of the Vulcanus's deadly cargo. Others, including the EPA's own scientists, are equally fearful of what could happen if the Vulcanus were to run aground, sink, or otherwise meet with calamity.

At the core of the concern of ship captains, scientists, and residents of the valley is the real fear, supported by both private and public research, that a spill of hundreds of thousands of tons of carcinogenic wastes in Gulf waters would create a major ecological disaster. Not only the marine life, but also the long-term health and commerce of those who live along the Gulf would be at grave risk.

The EPA response to this real fear is that there are contingency plans should a spill occur. However, the plans were drawn up only after the public expressed concern that no such plans existed. The U.S. Coast Guard is the agency primarily responsible for the cleanup.

I watched the Coast Guard "contain" the oil spill from the Ixtoc I well as it polluted the beach in South Texas. The twentieth-century technology employed by the Coast Guard and its subcontractors was shovels, sweat and straw. Hired hands load the contaminated sand onto the beds of dump trucks, which then haul it away.

As for "containing" a major spill of hazardous wastes in the Gulf, or any other ocean, the technology simply does not exist. There have been some good ideas on the drawing board, and perhaps one such will in the future meet with some success, but the Coast Guard has no magical method to clean up a major spill.

So why not keep all the sludge on land, where it is now, and dispose of it in land-based sites, as we do now? The at-sea incineration companies (CWM has been joined by three other newcomers) answer this question by demonstrating reduced costs in disposal. At-sea incineration will, it is estimated, save twenty cents on the pound by employing the new technology. When we consider the amount of wastes to be destroyed, it is clear that the savings would be substantial.

But I ask myself the question, "Who would see the savings?" Certainly not the public, because the cost of the inevitable spill will be astronomical. The general public, of course, will pay the bill through higher taxes, insurance premiums, or some other as-yet-undesignated assessment. In fact, this issue has become, at last, a national concern in a growing debate over the amount of liability the toxic disposer should be required by law to carry. As of this writing, those representing the public's interest are arguing that five hundred million dollars in coverage is the very least amount that should be required, while the representatives of the toxic disposal industry reply that fifty million is more than adequate to cover any potential damages.

At-sea incineration makes economic sense only if you are CWM or one of the other corporations involved in hazardous-waste disposal. Their gain is the public's loss. The motivation of these companies is clear. CWM has already earned an estimated $15 million from two previous test burnings of waste material in the Gulf. CWM, or whatever corporation eventually receives the permit, will earn approximately $250 million in contracts over a three-year period, and larger monies in the future for similar services.

Maybe I have misjudged this new technology, finding it inadequate more because of the failings of those who promote it than because of flaws in the technology itself. The disposal of wastes at sea has been carried out in Europe for more than a decade. At-sea incineration may prove a viable alternative, but we need time to gauge its advantages and disadvantages, time to make an informed decision. In the meantime, those who stand most to gain by its use are shoving the technology down our throats.

On the other hand, as resistance to this new handy technology grows, events that take place are far from reassuring. Rita Lavelle goes to prison. Block and Scarpitti, in their book *Poisoning for Profit*, disturbingly document the ties between major disposers of toxic waste, including CWM, and organized crime. Perhaps this technology can save us, but how can we be sure that those who employ it will be properly regulated?

Larger questions have gone unasked. Why, for example, does our society continue to produce toxic wastes at such an alarming rate? If we really understood the problems associated with destroying the junk, perhaps we would be far more willing to do without the benefits that make the toxic wastes necessary. It would seem far from absurd to recommend that we sacrifice a small measure of our quality of life if it would enable us to produce fewer harmful substances. But again, these issues are not the ones that, at a national level, let alone a state or regional level, are being discussed. There are choices. We can reject a proposed technological advance if it threatens us. We are not bound to accept every new toy.

The permits that CWM requested have been temporarily postponed by the EPA. Regulations to govern the burning of hazardous waste in the Gulf of Mexico are, at this writing, being finalized, even as the EPA's own Scientific Advisory Board has called for more basic research on various aspects of at-sea incineration. Burns of large amounts of carcinogenic wastes could resume, according to the EPA, by late 1985 or early 1986. Other sites off the East and West coasts are now designated, and hearings are being held. A bill recently introduced at the national level calls for a three-year moratorium on the new technology.

In spite of more recent attempts to arouse national concern, the public debate over land-based versus ocean-based incineration remains a local issue. Only those who have stared at the grim possibilities have become angered. CWM's efforts, and those of the other major disposers of toxic wastes, to minimize public debate and allay fears have been substantial. This is to be expected. There is a quarter of a billion dollars at stake. The leper ship is ready to set sail.

Winter on the Beach

A half-moon hangs behind partial clouds, white and cold backlighting for an empty beach. I can clearly see the gulls working the incoming tide. They cast half-shadows against the sand, their shapes fuzzy in the moonlight. Smaller, darker birds with spoon bills run between the gulls, in and out of the waves. I rub my hands together to keep warm. A mile out to sea I glimpse the red and green lights of a shrimp trawler, running upwind. The coast is devoid of beach walkers. It is the first week in December, a Tuesday, and the world is girding itself for the Christmas holidays. The beach is never better than in winter.

The beaches clear after Labor Day. The start of school draws families back to their neighborhoods, away from the summer beaches. Those who are left are childless, single, retired, or the recipients of untimely vacations. Yet fall at the beach has its rewards. The summer warmth lingers through the end of September and sometimes well into November. In October the waves often are warmer than the air; a wader or swimmer feels the chill only when getting out of the salt water and huddling between the folds of a towel.

The near-empty beaches that stretch for mile after mile inexorably draw me to them. Solitary beach walkers crossing paths again smile at each other, and comments about weather and tide are offered between strangers free now of the fear of conversing with a potential mass murderer. Even the gulls relax, no longer striving to hear the tune of a free meal. There is no food except what they capture from the sand and water. They grow leaner, temporarily free of Wonder Bread and cookie calories.

The crabs build their little circles in the sand, the circles that surround their holes. They stand by their escape routes but, seemingly aware that the children who pursue them have disappeared, make no immediate effort to sidle sideways down into their caves. Climbing the seashells and clumps of grass that are no longer gleaned by beachgoers,

the crabs peer into the distance, alert for signs of their natural preda-
tors. They sense they need not fear ending up in a child's sand bucket
or between the jaws of the family pet.

The jellyfish arrive in the late fall. They are swept up on the beach
by the thousands, their tentacles stripped away; then they die in the
warm afternoon sun, their air-filled purple bodies creaking and popping in the wind. During the daylight hours their bodies and bent tentacles are carefully skirted in case their sting somehow still remains.
But at night, under the stars, they are impossible to avoid, and walkers
without light crunch and crackle their way up and down the beach. The
popping of a jellyfish is like the cracking of a peanut shell. It is a minor
pop, but always an unforeseen one, and the autonomic nervous system
sends the leg jerking back in a spasm, as if the foot had been exposed to
a cruising shark. I crackle and crunch my way down the beach, thankful
when I reach an empty patch of sand, only to resume popping the purple air sacks that litter the beaches for miles.

The late fall and winter surf fishermen are serious about their
business. They stand in chest-high waders, bracing themselves against
the breakers. The tips of their long poles waver in the wind. When the
occasional blue crab or whiting passes by, the rods bend dramatically,
as if tied to a one-hundred-pound drum or jewfish.

The fishermen fortify themselves with thermoses of coffee, an occasional cigarette, and the sound of the waves. I stop to talk to them.
Most are marathon talkers, and I enter into their conversation knowing
they are willing to go the distance. Their talk is filled with the sights
and sounds of the beach. Once past the fishing report, they will go on
for some time about what they have seen, felt, and heard along the
beach. In their unhurried conversation is an appreciation for salt water
and wind, an appreciation that comes from standing hour after hour in
the surf with only one's thoughts for company. I admire their patience.

The fishermen disappear as the sun goes down; only a solitary fa-
natic remains, trying to get in a few more minutes of fishing before
total darkness. It is then that the trawler lights of the Gulf shrimpers
take on real meaning. They are out there in the Gulf, in five- to seven-
foot chop, backed by forty degrees Fahrenheit and the fifteen- to
twenty-mile-per-hour gusts. Under scant artificial light they seek the
elusive shrimp. I stand, arm level in front of my eyes, and measure the
pitch of their small vessels against the waves. Their lights jerk up and

down in the rock of the water. I know well their work, and the knowledge makes me uneasy. To the exhaustion and danger in a shrimper's work is added the bitterness of Texas winters.

Later, warming up in a handy beach house or condo, or in my favorite restaurant, I enjoy the exhilaration of having been wet and cold. For me it is the same exhilaration that comes after a good cross-country ski or a solitary thirty minutes of skating on a frozen pond. Warming up, one realizes how uncomfortable the cold was. Over a cup of hot coffee I complete the ritual of cleaning my glasses. The spray has laid a fine coat of salt against the lenses. As it accumulates over the duration of my walk, my sight is stopped down, like the shutter of a camera. As with the cold and the wet, I don't really notice the changes until they are rectified. Through clean lenses I peer around with a new curiosity. Later, under a hot shower, the film of grit and spray washes away. What remains feels cleaner than before.

The Christmas holiday brings a smattering of beachgoers who disappear when schools and colleges start up in early January. The weather grows colder. The jellyfish bodies disappear. The wind takes over. It blows down the beaches from a northeasterly direction, bringing with it the remnants of the frigid Canadian air masses. The gulls crowd together, backs to the wind, tail feathers ruffled by the blow of the northeasters.

The winds lift the sand and blow it down the beaches. At knee level it rattles your pants. The sand coagulates in the laces of shoes and fills the cracks between soles and uppers, in those hard-to-reach places that shoe polish never quite stains. With the wind at my back, I am pushed down the expanse of beach at a pace slightly faster than is comfortable. To compensate, I must lean back against the wind, shifting my weight to my heels. As I return, walking into the wind, the cold reddens my exposed hands and face as the gusts make walking upright an exercise in agility. I always start downwind to warm up, then figure that the walk back will take twice the energy and the time. Occasionally I grow frustrated with the effort, crossing over from the beach to a highway or road that offers some protection from the wind.

In January and February the beach condos and stilt houses lie empty except for those sheltering the few full-time residents. I walk past the high-rise ghost towns of steel, concrete, and stucco, only an occasional window illuminated on any one floor. Those who do reside within are warming themselves in the glow of their television sets. Outside the wind howls, the ground fog limits visibility to a few hundred yards, and the beaches are empty. The gulls have temporarily disappeared. I wonder what I am doing out on the sand, then realize that I am here in part because no one else is. The wind roars away at the sand and water.

The months of January and February are also good times to sit and look at the beaches and bays, times for remembering previous sailing trips, fishing excursions, picnics at the beach. Inside, behind a plate-glass window, the rays of sun warm my memories, softening the exact remembrances. The beach sand looks stark and hollow. I think back to its hot burn as I ran out of the dunes toward the surf, and I savor once again the sensation of liquid against burning toes as the waves wash against my ankles.

At my favorite bar on the bay I sit outside, ensconced in ski sweater and full parka, watching the February sun set behind the mainland. The waitress takes me for a loonie from the far north, immune to cold. I bus my own drinks. The small beach where I put my boat in has been packed hard as concrete by a winter storm. From my chair I can see the piling that keeps threatening to smash my bow to pieces. It lies almost totally submerged under the high winter tide. By summer its top two feet, now encrusted with oyster shells, will be black, oily, and iced with the leavings of the gulls. For now it reminds me of the last argument I had with it, when the boom came around faster than anticipated and I came, once more, within inches of its barnacles.

I love that piling. I have rooted for it and seen its victories over other men. Once I watched three men try to remove it. They spent a full day and, of course, failed. They used thick ropes, then chains, then dug at its foundation, cursing it as they went. They yanked at it with a powerful motor boat, then switched to a car bumper and, eventually, the rear end of a trusty pickup truck. Each time the weakest link gave way, either a rope, a fitting, or, in the end, the rear axle of the truck. They gave up, gathered all their equipment, then drove away. I count on that piling being there from one summer to the next, and to make sure it is, I periodically check on it during the winter.

From inside, a winter tourist appears briefly, camera in tow, snaps a picture of the setting sun, then melts back inside the bar. The waitress knocks on the inside of the window, making a motion to see if I want a refill. I nod and head back inside to pick up the drink.

Late February and early March are the last empty months on the beach. Soon the students will come, with frisbees, beer cans, and T-shirts. Through March and April they will catch brief glimpses of the beaches as the sand warms for the summer family crowds. I enjoy the summers, look forward to them, but I thrive during the Texas winters on the long walks along the empty, endless beaches.

Broil It

I grew up in central Oklahoma in the 1960s, when fish sticks were in fashion. They were almost the only fish I knew. In my teenage years my mother fried them up, and I ate them by the dozen, with a pound of catsup on the side.

Our house in the suburbs was a little different from the rest. It bordered a large field and a land-filled area harboring a drainage ditch that ran from the street in front of our house under the ground to a spot about fifty yards beyond our fence. The ditch ended in a sluggish backwash that, at its summer height (less than a foot), turned a brilliant green. In that backwash I stalked crawfish with spear, net, and brick, occasionally bringing the crustaceans home to grow stale in jars stored in our garage.

During a wet year the ditch also housed minnows that collaborated with the crawfish. They were much harder to catch, being insensitive to spear and brick; for every ten crawfish captured there was only one lone minnow, brownish, with slightly protruding eyes. Although the crawfish seemed to grow larger during the summer months, the minnows reached their hopeless maturity early, soon disappearing into the mud bottom of the stagnant waters.

The fish sticks my mother cooked were only slightly larger than the minnows. I liked them in part because we could eat them with our hands, like fried chicken. I remember the smell of their cooking in the kitchen, along with the occasional shrimp, also fried. I was apportioned a certain number of shrimp as a child, and as I grew, so did the number. By the time I was sixteen I was up to seven fried shrimp, plus any left over after everyone else had finished. I used to count them as I dunked them in the catsup. Fish sticks and shrimp, minnows and crawfish, they all were a part of my childhood amid the dry, rolling plains of central Oklahoma.

The only other fish I ever came across during those years were perch found in a local pond. There was a fishing tournament held every year at the local country club, and I entered a few times. I actually remember the catfish better than the perch; even then they had a bad reputation. One of the judges at the tournament would remove any catfish he saw and, with a grimace, fling it back into the water. Perch were trophy fish; catfish were sewage eaters.

Most of my fishing time growing up was spent in pursuit of what we called "clams." These were fresh-water mollusks unfortunate enough to populate a pond within easy walking distance of where my parents took us to swim during the summer months. My friends and I often grew bored with the swimming pool and the hot dogs and, by early July, had perfected our shellfish hunting equipment to the point that we rarely returned empty-handed.

The clams hid out in two to three feet of water not five feet from the bank of the pond. After running through various arrangements made of sticks, tape, and wire, I finally hit upon the perfect Oklahoma clam contraption. My mother had bought a set of four hot dog sticks, the kind that were made of thick black wire, thicker than coat hangers, and which at one end open and shut to hold fast the hot dogs. They were about three feet long, all of it handle, and were lying behind an old table in the garage wrapped up, all four of them, in a rope.

After several adaptations and false starts, I and occasional accomplices began systematically to deplete the population of mollusks in central Oklahoma. With my new invention, failure was a thing of the past; leaning out over the water, I would scoop up an unwary mollusk that had buried itself in the mud. After washing it off, I tossed it in a metal bucket for the car ride homeward and the eventual safekeeping in our garage. The method was foolproof; even my worst enemies received clam gifts that were quickly relegated by their parents to the garage.

Gradually, however, a serious problem arose. The clams seemed to prefer ponds to pails; rotting shellfish in airless summer garages offended an increasing number of parents. After all, little could be done with these clams. Like the crawfish before them, once captured they were subject to long-term neglect. The heyday of my clam hunting grew to a close as the social pressure mounted to dispose of the bodies. I returned the hot-dog roasters to their dusty place in the back of our garage.

I imagine that my limited experience with fish and related species, both the eating and the hunting of them, is shared to some extent by many who grew up in cities and towns some distance from lakes, rivers, or other large bodies of water. Our attitudes toward fish and fishing are, for the most part, the sum total of what we learned while growing up, mitigated by what we learn as adults from others, whether we live north or south of the Red River.

Texans, like most Oklahomans, never have cared much for eating fish, any fish, but have thoroughly enjoyed hunting fish down. When Texans placed their prize catch on the kitchen counter, many of their wives grimaced and returned the gift. Texans, both male and female, eventually took the easy way out; they threw the fish into the nearest pan and fried away. And Texans are still frying fish, regardless of freshness or species, from Port Arthur to Brownsville. The trend has culminated, more recently, in the ubiquitous Cajun dish of blackened this or that. I admit that the first three bites of blackened redfish, for instance, are mouthwatering, but I've never yet been able to finish this Cajun version of fried fish.

Food, of course, is but a part of our larger culture. We tend to eat what others around us are eating. While fish eating is on the rise in this country and in Texas, it still comes nowhere near our per-capita consumption of beef. Americans consume, on the average, about thirteen pounds of fish a year, compared with more than a hundred pounds of meat on the hoof. Most of us still prefer a quick Quarter-Pounder with fries to the fish from the franchised restaurant across the street.

The fact that we fry our fish demonstrates to me that we have an even longer way to go than societies that have long since discovered the pleasures of eating broiled, baked, or even raw fish. There is, of course, a strong nutritional argument for not frying fish, or any foods, for that matter. Simply put, fried foods have more of the wrong kind of calories. Doctors advise people with certain kinds of health problems, including heart disease, to stay away from fried foods. But I think there is an even more convincing reason, and perhaps a more practical and compelling reason, to stop cooking fish in liquid fat. Fish tastes better if you don't fry it.

I never asked my mother what kind of fish went into those fish sticks I consumed by the ton. Fish was fish, whether perch, haddock, or trout. Regardless of what species is cooked up, frying reduces it to the lowest common denominator. Fried fish becomes generic fish, without identification. Blending delicate flavors of certain fishes with hot grease destroys their distinctiveness. What is left, according to my own taste buds, is another fried Whopper—not that it is bad, just that it is much less than it could be.

Broiled and baked fish retain more of their original flavors and/or the flavors the cook adds to them. To those who would quickly reply that it is exactly those fishy tastes they are trying to get away from in the first place by frying the fish, let me again emphasize that we enjoy food, any food, only because we have acquired a taste encouraged by the particular culture in which we live. Take scotch, raspberry parfaits, and okra as three examples. One's bias against nonfried fish, then, can change.

The difference between fried and, for instance, broiled fish is not just the difference between a chicken-fried steak cum gravy and a Texas T-bone with a cool center. All fish does not taste the same. There are some species I prefer, such as flounder, sea trout, redfish, and shark, and some I would just as soon had not been invented. Salmon, for instance, leaves me cold, although if it is turned into lox, slapped over a hot bagel with some cream cheese, that is a different story.

The other problem with eating fish, whether it is fried or not, I have already mentioned: it continues to be compared unfavorably with American beef. Somehow to suggest one likes fish is to be less than patriotic, especially here in Texas, where our cultural history is dominated by the pervasive image of cowboy and rancher. But the two meats are not incompatible. One need not be either a beef or a fish eater, the two species demanding no loyalty in and of themselves beyond what our culture instills in us. All things being equal, if some of our cowboys and ranchers ate more nonfried fish, they would probably live longer to punch more dogies. So would our commercial fishermen, for that matter, who consume as much beef as other Texans, regardless of their familiarity with fresh fish.

Of course there will always be those who do not prefer any kind of fish—fried, nonfried, or petrified. When they think of fish, they probably recall smells similar to those of the Oklahoma clams I stored in my garage. So give me their share of fresh red snapper, the ones that go between a pound and a pound and a half, broiled over a flame, lightly seasoned with lemon and butter, and they can criticize fish eating and fish eaters all they want.

After the Hurricane

Hurricane Allen blasted into open country slightly north of Port Mansfield on August 9, 1980. The small community was submerged for more than twenty-four hours and buffeted by winds of more than one hundred miles per hour. Hurricane Allen left in its wake yachts planted on the highway, houses and trailers blown or washed off their foundations, and tons of hurricane flotsam sprinkled around the community: glass, concrete fragments, boat parts and ornaments, children's toys, plastic bottles, all the paraphernalia of modern times stirred up by high winds and saturated with salt water. Many residents of Port Mansfield now say Hurricane Allen was the best thing that ever happened to them.

I drove up to see the effects of the hurricane almost as soon as the town was reopened to outsiders by the National Guard. Residents were going about their business cleaning up, rearranging and moving their household possessions, fixing their roofs, painting, sawing, helping each other. A closer inspection revealed missing pieces of road, buildings shifted to odd angles, and boats and boat pieces in fields across the street from the public marina.

Five years later Port Mansfield is a different community. Still divided by geography into three sections—the north side, the harbor, and the south side—Port Mansfield has experienced a rapid increase in the number of private residences and condos, an influx of marine-related industries, and an expansion of trailer parks, retail stores, and public facilities. Although still boasting only 177 registered voters, Port Mansfield is booming.

The north side now bustles, relatively speaking, with more than a hundred summer cottages and year-round residences, the majority perched on stilts that face the bay waters of the Laguna Madre and, eight miles due east, Padre Island and the Gulf beyond. Most of the houses are brand new or sport new additions and remodeling jobs. There are boats, big expensive fishing boats on trailers, in the driveways or under the pilings of every other house. Wooden fishing piers extend into the shallow bay waters. At their ends are covered areas to protect fishermen from the winds that blow constantly off the bay.

A few streets behind the stilt houses and the occasional ground-floor dwellings are two trailer parks, completely filled, bumper to bumper, from November to the first warm days of spring. Slightly farther to the west lies a lighted 3,200-foot runway, big enough for executive jets. A new shrimp farm is to be constructed north of the airport, turning 150 acres of salt flat into productive land. Bayside, the Chamber of Commerce building sits at the end of the road, alone and deserted during the day but at night the hub of the town's social life. Across the street is a small park complete with concrete cabanas and picnic tables. Fifty yards farther north is a barbed-wire fence, the town's northern boundary and the beginning of cattle country.

Along newly widened Highway 186, which leads to Raymondville, twenty-five miles to the west, is a new little park with tennis courts and swings. Across the street is the harbor area, which now sports covered boat stalls, concrete pilings, and new wooden docks. On the south side five new condominium developments have been constructed since the hurricane, and one more is under construction. Four oil-supply companies have moved into spaces on the north side of the harbor, and the muffled roar of their huge diesel supply vessels can be heard around the clock. Only the empty fish-packing plant next to one of the marinas belies the obvious growth and prosperity.

It is almost a two-mile drive around the harbor area, from the north side of town to the south, and the transition from upper middle class to white working class is evident. On the south side children's toys dot the front yards of brush and grass, and there is human noise in the air; on the north side the streets and yards are abandoned most days of the year, except summers and weekends in the fall and winter.

The south side, the original townsite, has been spruced up since the hurricane. Although a few ramshackle buildings seem to be standing by sheer will, it is easy to see in what direction the neighborhood is moving. The less affluent live here—not the condo owners from Houston or Dallas, who live in $125,000 dwellings, but the full-time residents, those who consider Port Mansfield their home. They work as guides, enforce the law, wait on tables, run laundries, clean condos, build new houses, and manage the summer residences. They have painted and fixed up the shacks, added on rooms, garages, and outbuildings, built fences around their yards and omnipresent fishing boats (more modest than those on the north side) and generally dug into their pieces of sand as only people do who call a place home. Up and down the sand streets, squeezed between the small houses, sit new $100,000 houses, the owners obviously preferring the south side because of its lower prices and its ambience. The money invested in Port Mansfield has trickled down, and those on the south side who have earned some of it display, in a more modest way, the new prosperity.

A new road south of town follows the spoil banks and leads to Port Mansfield's new and only suburb, Port South. To date Port South is little more than a sign, a road, and five expensive houses on stilts. But the lots are selling fast, and the development, directly on the Laguna Madre, offers a place for those who want to live like those on the north side of town but cannot find a lot. The five houses, priced in the $100,000 to $125,000 range, stand starkly against the brush and miles of bay extending south to South Padre Island. If Port Mansfield continues to grow, the structures will soon blend in with the other hundred-odd houses that will jam the new beach-front lots.

According to the people in Port Mansfield who should know, the unprecedented growth began immediately after the wind and water of Hurricane Allen had receded. The community, originally founded in the 1940s to serve as a supply port—the only one between Corpus Christi and Port Isabel for the Texas shrimp fleet—was for many years nothing but a haven for dedicated weekend fishermen and the home of shrimpers and commercial bay fishermen. The Willacy County Navigation District, which actually owns the 1,700 acres that make up the community, leases the lots to individuals and companies. The Navigation District, most residents agree, has provided the leadership and expertise that has fueled the recent prosperity.

The director of the Navigation District points to the private and federal dollars pumped into the community after the hurricane. According to him and others, private investors feel that Port Mansfield, having weathered a major hurricane, can now look forward to at least ten years of calm summers and falls. Whether this is true or not is less important than their *belief* that Port Mansfield is a safe place to build. Nor is this rationalization unprecedented; just down the coast South Padre Island underwent rapid development in the late sixties after Hurricane Beulah.

Hurricane Allen drew national attention to tiny Port Mansfield.

Investors, predominantly from Dallas, Houston, and San Antonio, heard about Port Mansfield, and the publicity gave the community an edge over the other developments and communities dotting the Texas coast. Now, serious weekend fishermen in particular could better justify, to themselves and their families, their substantial investments in housing. The houses, perched high on pilings, will not be bested in the winds of a hurricane for some time to come; for now Port Mansfield is, according to popular thinking, virtually hurricane free.

Besides the improbable belief that storms will leave it alone, Port Mansfield differs from the majority of other small communities up and down the coast in the way rapid growth there has been rigidly controlled. In the director's office in Raymondville a color-coded map depicts very clearly what is happening to Port Mansfield. Through strict zoning regulations the community has steered clear of the haphazard growth that characterizes communities such as Port Isabel, where one finds a mobile home next to a condo and across the street from a discount store. In many other coastal communities, variances—permissions to circumvent the building code, most notably including constructions in a flood plain or directly on the sand—are often bought and sold from the local zoning commissions.

Residential dwellings in Port Mansfield have dramatically appreciated during the last four years because there is a limited number of desirable lots in the north and south sides. In 1979 a condo unit alongside the harbor sold for $40,000; now, because the space for waterfront condos is severely restricted, the same unit sells for more than $100,000. The last units in the harbor area will be finished early in 1985. Likewise, industry is confined to certain sites along the harbor and other restricted areas in the community. The new shrimp farm, for instance, is virtually hidden from view on land undesirable for anything other than raising shrimp.

By holding to the zoning codes, by sticking to lease restrictions on houses, and by keeping an eye on industry, the Navigation District has ensured that growth in Port Mansfield will be orderly and contained. Not only are the natural quality and beauty of the environment protected, an asset that in and of itself attracts new growth, but land and water resources (in limited supply all along the Texas Coast) are wisely used as well. The harbor itself reflects the multiple applications of available resources. On one side men busily load oil-supply boats between a motel and a recreational boat marina; on the other, condos nestle not far from two restaurants, a gift shop, covered boat docks, a commercial fish house, and the Coast Guard station. Offhand I can think of no other Texas port, large or small, in which water, shoreline, fishery and mineral resources, wholesale and retail operations, and public recreation facilities are in such harmony, the users working together instead of stepping on each other's toes.

More important, the community leaders seem to have an idea of what they want Port Mansfield to be in the future. Such foresight is almost unheard of along the coast. Leaders who make decisions that will affect the growth of this community in years to come seem committed to the enforcement of zoning ordinances. They will not authorize construction on floodplains, grant large numbers of variances, construct ten-story high rises that obstruct the view, or find ways to circumvent the regulations of the Environmental Protection Agency. If new industry comes (it is certainly welcome), it will have to follow the rules. As a group these men and women seem to realize that it is in their own best interest to keep Port Mansfield highly regulated. Their decisions have paid and will pay off in dollars.

There is, however, a darker side to this prosperity. For a long time Port Mansfield was a haven for commercial fishermen, who, with their families, subsisted on the bay fishery, especially on trout and redfish. Now, however, the rapid increase in property values is having one of two repercussions on commercial fishermen. Either the increase in rents and other services is simply driving these people away, or it is forcing them into new jobs. A major impetus behind this shift is the recent legislation banning the commercial fishing of trout and redfish in Texas. But if the ban had not forced people of limited means to look elsewhere for jobs, sooner or later the cost of living in Port Mansfield would have done so. Even on the south side, a small cottage that sold for $20,000 before the hurricane now goes for double or triple that price.

Some of the commercial fishermen in Port Mansfield left town after legislation was passed banning the use of gill nets in the bay. Those that did not get their licenses as fishing guides (nearly forty men have them) are working in construction, on the oil boats, or at whatever they can find that pays the rent.

Low-income families from San Perlita, Raymondville, Harlingen,

and other valley towns, along with middle-income families, are finding
that Port Mansfield both hits their pocketbooks harder and offers less
than it did a year earlier. The new waterfront condos now form a solid
chain of buildings and fences along the south side of the harbor; a
weekend angler can no longer fish from the piers without trespassing
on someone's expensive property. On the north side of town private
piers bristle with Keep Out signs. Prices in the local restaurants rival
those of South Padre Island. Motel rooms in one lodge, for instance,
are forty dollars during the off season, fifty-five to sixty during the fishing season. It costs more and more to fish in Port Mansfield.

A brand-new, members-only yacht club is being constructed on the
harbor next to the old Redfish Motel. Initial membership is $20,000,
monthly fees $200. At the county park, six blocks to the north, Mexi-
can-American families take advantage of the only free shelters and
grills in town. They eat their Friday dinner out of big green coolers.
The men fish from the only free public fishing pier on the bay. Most of
the women sit at the tables or in the cabs of their trucks conversing
with their friends. Their children run up and down the long pier, from
fathers to mothers and back again.

The two trailer parks are filled with those who cannot afford to buy
a summer house or pay the $550 a month for a small two-bedroom
house. During the winter months the parks are filled with retirees,
snowbirds, from the north. The summer months bring the serious fishermen and their families with the family boat. Sometimes the boat is
better cared for than the small family trailer. The trailer-park occupants save money by avoiding the condo and motel prices as well as the
high costs at the restaurants. They cook most of their own food, food
bought in Raymondville at the HEB or brought with them from home.
The irony is that the planning that preserves Port Mansfield, that protects it from many problems facing other communities up and down
the Texas Coast, has made the town both more desirable and, at the
same time, increasingly inaccessible.

Prices have also risen dramatically because local real-estate interests and developers are working to transfer the prices of real estate
from other coastal markets to Port Mansfield. Port South is an excellent example. The waterfront lots of the new development, selling at
$25,000 apiece, were gone in a year's time. The majority of those who
purchased lots invested their money not just in what Port Mansfield is

today, but in what it would be five or ten years from now. These buyers are less interested in building a home on their property than in betting that several years from now they can sell their lot at a nice profit. Speculative capital lured by local developers helps drive up all real-estate prices. It is turning the community into a much more expensive place in which to fish and live. Should even more offshore fishermen and their families decide to move to Port Mansfield from their favorite spot, Port O'Connor, prices will rise still higher.

At the local bar, the Windjammer, talk occasionally turns from the citification of Port Mansfield to the good old days before Hurricane Allen. Someone always points out that no problems in this community by any stretch of the imagination compare to those of urbanized life in other Texas metropolitan areas. Even so, during my last stay in Port Mansfield I was surprised to see on the wall of the Windjammer two impressive renderings by an architectural firm. Hanging directly across from the cash register, the drawings depicted various views of the new condo units going up on the very ground on which I stood. Plans call for the Windjammer Restaurant and Bar to become the Windjammer Condominiums, complete with pool, sauna, boat docks, sundeck, and units with living room, dining room, kitchen, and two to three bedrooms. The land is becoming too expensive for a bar and restaurant; the locals will have to go elsewhere to talk about the good old days.

In twenty years Port Mansfield is still going to be a pleasant place to live, barring any direct hits by a hurricane. Its bay waters will be clear, its residences well maintained, its industry clean and productive, and its people friendly although more numerous. Through continued abuse of coastal resources some communities and developments along the Texas coast will become coastal slums or ghost towns. Port Mansfield will grow old gracefully; it is one of those rare places in Texas where fishing will always be good. But if you want to hold a rod in your hands, sit and watch a sunset, or otherwise enjoy what this community has to offer, bring plenty of money.

A Modest Proposal

The more I've lived and traveled along the Texas Coast, the more questions I've asked about its past, its present, and its future. Many of these questions, these issues, I have discussed in the previous pages. The partial answers I have discovered concern me as a researcher, a father, a Texan.

Twenty years down the road we have taken, those who are rich will still be able to afford their small piece of the Texas Coast, a plot relatively clean and unpolluted. The rest of us may well be stuck with coastal lands that approximate those surrounding the New Jersey Turnpike. Many Texans who have lived along the Gulf of Mexico in the same ways as did their grandparents before them will be displaced; they will catch a glimpse of the setting sun, not from their homes on the bay or Gulf waters, but from several miles inland, where they can still afford the prices.

The crucial problems, though sometimes complex, are not many. Simply put, we are applying increased pressure on limited coastal resources, on land, water, air. For most of the human history of these lands, no more than ten thousand people depended for their survival on what this vast area could offer. Today several coastal counties support millions. Other county lands are routinely misused. Even as we build new projects on sand that borders the Gulf—new homes, banks, restaurants, convenience stores—we are both fouling the surrounding bays, estuaries, and rivers, and depleting other available resources.

At the same time, many long-time residents are being driven off by the economics of real estate. The land on which they live has become too expensive for them. Those fortunate enough to own their homes can benefit, at least in the short run, from increases in property values. But the lower economic classes, the majority, cannot. Even as their homes become too expensive, their jobs disappear. Many commercial fishermen along the Texas Coast have found themselves no longer able to fish "legally"; many others, notably the Gulf shrimpers, are victims of national and international economic trends.

The pressure we put on some coastal lands is not new. Reports of serious contamination go back at least to the nineteen twenties, when bathers on Galveston beaches were forced to use cans of gasoline to remove oil sludge from their bodies. Even then the Houston Ship Channel was judged among the most polluted areas in all the United States. Now, within the last ten years, as some coastal counties have undergone rapid growth and development, the pressures on limited coastal resources have intensified. This is not to say that all coastal industries consciously pollute, or that all developers go out of their way to damage the local ecology. Rather, it is to observe that, given the present state of affairs, those that do are often ignored and those that do not are rarely praised.

The problem, in part, is one of regulating and managing what we still have. The history of the protection of the Texas Coast by the designated regulatory agencies is inconsistent and sometimes sordid. There has been a continuing conflict between some in industry who would put their short-term profits ahead of the long-term welfare of the general public. Over the last eighty years city, county, and state agencies have attempted to limit the more obvious abuses of the coastal environment. More recently federal agencies have lent a hand. Yet the problem remains: Texas has no plan to protect existing coastal resources and thereby provide for the future. In some ways it is still much easier to break existing laws than to abide by them. The Texas Coastal and Marine Council, which had provided some leadership on certain issues in the past, was abolished as of August 31, 1985.

We need a plan. We need to determine objectives and goals, protect what is still there. Various levels of government, often understaffed, have won specific battles but are losing the war. Fortunately, there are all kinds of models to choose from. Florida, for example, has a state management plan that, although not perfect, is a big step in the right direction. The plan outlines state priorities along its coastline, how it will implement them, and who will enforce the regulations. It is not antibusiness by any means; rather, its goal is to encourage business growth along with the wise use of existing resources.

Here in Texas, however, we seem to reinvent the wheel every time there is a coastal crisis. In fact, we seem to go from one crisis to another, from one oil spill to another, one hurricane to another, one variance to another. And each time city, county, state, or federal regulations are hauled out and used as best they can be—generally ineffectually.

The recent spill by the tanker *Alvenus* is a case in point. A task force made specific recommendations in a seventy-three-page report. But because there is no state plan and no state agency appointed to enforce the plan or the suggested regulations, the recommendations have little chance of ever being implemented. Brazoria County now has in place, in fact, a plan to respond to an emergency oil spill. But it remains the only Texas county that is ready to react to such a disaster.

A plan, of course, is just the first step. No plan or new state agency can by itself reverse the downward slide of the coastal environment without base line data about the coast itself. We have many bits and pieces but no broad analysis. As a result, not even the experts agree about the seriousness of the problem. Until we know the current status of the coast, it will be difficult to judge, let alone measure, changes when they do occur.

A plan and a status report are beginnings. There is also the complex problem of how to handle the international aspects of control over our Texas Coast, whether involving an oil spill by an unlicensed foreign captain operating in Gulf waters or systematic incompetence by a corporation with holdings around the world. To help us tackle the larger issues, we need a plan and some stiff enforcement.

And what of the residents of coastal communities? Sometimes in a rush to save the environment we forget about those who depend on it for their living. How can local community residents protect their own interests? How long can they continue as fishermen, or in jobs directly related to the tourist industry, when the lands and waters on which they live and work are a little more polluted each day?

There are no easy answers. It is too much to expect that rural, traditional peoples will suddenly define their community interests as distinct from the interests of those who stand to benefit by rapid change. And the case is far from being clear cut. Some residents benefit from rapid development in a particular community; some do not. Certainly the creation of regional organizations that focus on issues pertinent to all the communities is one way in which many long-time inhabitants of the coast could begin to address the problems they, and others like them, face. There are choices to be made. The ways in which specific communities change and are changed can in no way be predetermined; nevertheless, the changes are the outcomes of deliberate decisions.

For my part I offer a modest proposal, a beginning. Let us push our legislators to create a management plan like those that other states have implemented with some success. Once in place, such a plan will provide guidelines for present and future uses of the Texas Coast. Among the planners, let us include those with vested interests in these lands, both the large petrochemical conglomerates and the small businesses that depend on coastal resources. Let us acknowledge that growth is desirable in coastal Texas, that a good business climate is a priority. Let the general public voice its opinions, too; let the various segments of the public incorporate their own vested interests in the plan. Once we have a solid plan, we have a chance.

Fixing Her Up

Moldy green sails, encrusted stainless steel fittings, she followed me home on rusting trailer wheels this fall. There, in the backyard, I placed her next to the garage, which has been falling down for five years. They were suitable companions, the sailboat and the garage, the one evoking guilt and the other annoyance. I had every intention of fixing up the boat, but I never seemed to get started. I thought bringing the boat back from the island would generate enough guilt in me, thought I could not ignore an eighteen-foot boat in my own backyard. As for the garage, the sole reason it remained standing after too many years of service was that some tired termites had quit before finishing the job; on my list of priorities, several notches below the sailboat, was the demolition of the garage and the building of a new one. The two, the sailboat and the garage, made quite a pair.

I covered the boat with a bright orange tarp, avoiding with my eyes the blatant signs of neglect from bow to stern, and fastened the tarp down with cords. My plan was to begin attacking the rust and corrosion that weekend. But something came up, I can't remember what, and the boat sat untouched through the weekend. In fact, she sat there for five months, enduring the big winter freeze that dropped pounds of ebony leaves, sour-orange branches, and unidentifiable objects that floated through my backyard. The surfaces of the boat unprotected by the tarp became layered in heavier-than-air vegetable matter that, with the winds, sifted between cracks and crevices and found a real home in the anchor locker. With the rains came new wildlife looking for a dry port. On the small side were the omnipresent cockroaches, who must have found slim pickings among the fiberglass, aluminum, chrome, and steel.

I ran a constant campaign against the spiders. They found the track in the mast ideal for hiding. One in particular, brown-bodied and double-sphered, was just the right size to fit with vulnerable organs inside the mast groove and legs dangling outside, next to his symmetrical web. Every two weeks or so I destroyed the web. Finally, in frustration, I mashed him, two globes and all, against my shoe. Rust I could tolerate; spider webs were just too much.

On the larger side, several neighborhood cats considered my boat a home. I chased them away. They returned. My own cat taunted me, too. She sat on the bow, amidst corrosion and litter of decaying leaves, delicately cleaning her paws. She contributed to the boat her own set of paw tracks, which led all around the small walkway. The neighborhood possum, who had grown rotund on our sour oranges and refuse from the garbage cans in the alley, no doubt also gave my boat the once-over. If it had come to that—my sailboat a home for possums—I was prepared to evict by force of a gun.

It all ended one Saturday in early March, more than five months after I had hauled her back home. Acting as if nothing whatsoever were out of the ordinary, I gathered my tool chest, steel wool, lubricants, extra hammer for leverage, and a big glass of Pepsi and headed for the backyard. I removed the tarp, hosed down the boat, and began to assess the damage. Then I spent fifteen minutes hopping around on the pavement by the boat removing ebony stickers from my feet. I washed off the concrete, cursed the inventor of a tree that served no useful purpose except to provide scattered shade during certain unimportant seasons, and attacked the boatly crud.

Beginning with the hardware and brightwork closest to the mast step, I slowly moved aftward. I stripped the cleats down to their springs, removing the screws, washers, and thin stainless steel plates that keep them in place. I cleaned the springs, wiping away what I could of the corrosion with a rag, using steel wool on what remained. The noon heat beat down on my back. Occasionally I reached for the glass of soda and ice, but my attention remained riveted on any small dot of rust that remained. I put a drop of oil on each spring, reassembled the simple mechanism, and attached it to the fiberglass hull. With my fingers I now flipped the overhauled cleat back and forth, enjoying the improvement in both its appearance and its internal workings. Where before it had been sluggish, slowed down by the warp of salt, it now emitted a healthy click when I ran a line through it.

I completely lost track of time. As soon as I cleaned one piece of equipment, I moved on to the next. The satisfaction was exhilarating. I stopped when I realized I needed something cold to drink and that my back was red from the sun. Throwing ice into my glass in the kitchen, I glanced at the clock on the stove. More than an hour and a half had passed, had glided by under a hot sun and the enjoyment of the work.

I hosed off the boat, gathered my tools, threw on the tarp, and went off to take a shower. The next weekend I was back, ready to pick up where I had left off. This time I had Travis to watch while Andrea tried to catch up on her sleep from the night before. Our baby, Lauren, does not believe in sleeping the night through except in two- to three-hour shifts, punctuating them with demands for food, entertainment, or both. I was not pleased at the prospect of working on the boat and watching after a four-year-old. The two activities seemed totally incompatible, but, on my wife's advice, I was going to give it a try.

I hosed down the boat, freeing it of a week's worth of litter, opened up my tool chest, and set to work on an encrusted fitting. Travis, ten feet away, climbed his jungle gym and stared at the boat. A minute later his head popped up over the bow, his eyes silently focusing on mine. I continued working, he continued staring, and the grackle in the ebony tree provided an asymmetrical sound track.

"Travis, if you want to work with Daddy on the boat, you have to climb up yourself."

"I know how to do it, Daddy."

"Okay, then do it." I felt confident that he would not be able to climb up over the bow, which rested on the trailer a good four feet off the ground. But if I told him he could not do it, he would get mad. By telling him he could, when I knew he could not, I reasoned he would try for a little while, get frustrated, then go off to play with his other toys. The convoluted wisdom of a relatively new father.

He grunted and groaned trying to raise himself over the bow, then was quiet. I thought he had given up. Suddenly he was sitting next to the anchor locker, big smile of victory on his lips, at the same time looking a little bewildered that he had made the climb. I am still not sure how he did it.

I held his hand while he half crawled, half walked his way to the cockpit. I had to keep him busy or in no time at all the boat would be a disaster area, like the entire back of our house, which we had ceded some time ago to our children. I gave him the hose, which I kept running at my feet, and told him to clean up the dirt on the back of the boat. He said he needed a big screwdriver to do the job, so I gave him one. Screwdriver in one hand, hose in the other, he staggered around the new environment, pleased with his responsibilities.

With Travis on board it took twice as long to accomplish the same amount of work. He constantly stopped to ask questions and make comments, and when he closed the bailers in the stern, the water backed up to my ankles before I realized what he had done. But he enjoyed helping, and I enjoyed sharing the boat with him, even if it made the work go more slowly. I told him that as soon as I got the boat all clean we would go sailing. He asked where the motor was. I told him sailboats do not use motors. Then he told me that we would take the wheels off the trailer and put a motor on the boat and go sailing. I said okay.

After that episode with Travis, working on the boat was far easier. The major boat-guilt disappeared. Slowly the boat was transformed back to sailing status. I found that steel wool removes not only rust, but also the yellowish ebony tree stains on fiberglass that a stiff brush cannot budge. All that now remains, in late April, are the fittings on the mast and the moldy sails. I am well on my way to knowing the boat, as each part of it begins to gleam, once again. This summer Travis and I will, together, sail the bay.

Epilogue

In mid-December, while the rest of the country freezes through one of the coldest winters anyone can remember, Andrea and I load our children, Travis and Lauren, and head for the beach. After a big breakfast at our favorite restaurant, we make a final decision to ignore those snug in their homes awaiting spring. The fog has miraculously cleared, a bright sun prevails, and the winds, from the northeast, have all but disappeared.

Mounting the dunes, my son on one arm, his beach equipment on the other, I see only a few beach walkers far down the sand. Travis runs through the litter that lies between the dunes and the beach, stops to view the surf gently lapping against the sand, then looks back toward me. I in turn look over my shoulder at Andrea, who with Lauren in arms is walking along the dune path. Lauren comes prepared. Wearing a slightly used diaper and an old T-shirt, she is ready to tackle her first sand castle. That morning she will also taste her first handful of sand, self-inflicted.

As I stand in the winter sand, I want many things for my children. I want their imagination to soar at the sights and sounds of the waves. I want their hearts to reach out into the marshes, bays, salt flats, brush, and dunes and understand the changes that are the essence of this land that borders a sea. I want them to have the foresight and knowledge to guide and protect these lands and waters for generations yet to come.

I want my children to have the chance to stand at sunset, silhouettes against the orange globe that disappears into the bay waters, and quietly say goodbye to another day.